Complex Systems

Fundamentals and Applications

Complex Systems

Fundamentals and Applications

Edited by

G. Rzevski
The Open University, UK

C.A. Brebbia
The Wessex Institute, UK

WIT*PRESS* Southampton, Boston

Editors:

G. Rzevski
The Open University, UK

C.A. Brebbia
The Wessex Institute, UK

Published by

WIT Press
Ashurst Lodge, Ashurst, Southampton, SO40 7AA, UK
Tel: 44 (0) 238 029 3223; Fax: 44 (0) 238 029 2853
E-Mail: witpress@witpress.com
http://www.witpress.com

For USA, Canada and Mexico

Computational Mechanics International
25 Bridge Street, Billerica, MA 01821, USA
Tel: 978 667 5841; Fax: 978 667 7582
E-Mail: infousa@witpress.com
http://www.witpress.com

British Library Cataloguing-in-Publication Data

A Catalogue record for this book is available
from the British Library

ISBN: 978-1-78466-155-7
eISBN: 978-1-78466-156-4

Library of Congress Catalog Card Number: 2015943218

Printed by Lightning Source, UK.

Preface

This Volume comprises a series of papers dealing with new approaches for resolving complex systems, a topic that has numerous applications in modern society.

The papers demonstrate how complex issues, which cannot be formulated using conventional mathematical or software models, can be resolved.

Complex systems occur in an infinite variety of problems, not only in the realm of physical sciences and engineering, but encompassing fields as diverse as the economy, environment, humanities, social and political sciences. This book demonstrates how complex problems can be solved in different disciplines.

The Editors hope that this work will promote further applications and interaction between researchers working in different fields, as the solution of complex systems nearly always require interdisciplinary approaches.

George Rzevski and Carlos A. Brebbia
Editors

The New Forest, UK

Contents

Section 2: Applications in business and industry

Section 3: Applications in various fields

Section 4: Multi-agent applications

Section 1
Fundamentals

Non-relativistic time, existence and adaptation

R. G. Boothroyd
Retired Chartered Engineer, Queensland, Australia

Abstract

By using non-dimensional Navier–Stokes equations in systems where partial modelling complications can be avoided, it is shown that geometrically similar systems can exhibit identical behaviour but according to different time scales where the latter are also within our own control. These models indicate that the physical nature of time is not related to a fundamental constant, but the passage of time can have a different but still constant value in a particular system of our own choosing. This observation also adds to our knowledge relating to the dichotomy of interpreting time either as a flowing parameter or, alternatively, just accepting it as a part of space-time. The observation adds evidence in favour of both viewpoints. The conclusions also seem relevant to biological species, chemical processes in general and other branches of physics. Using earlier work of Prigogine and others, it also appears that instantaneous time has important properties dividing past and future time into two segments which differ fundamentally in physical characteristics. This seems to explain the 'raison d'tre' of the second law of thermodynamics and the necessarily asymmetrical nature of time in our own world where mathematical symmetry of physical laws is the norm. This entropic time barrier also seems to explain the physically intangible nature of past events and their possible natural importance as former lower entropy forms of existence. It is concluded that these physical features of time are necessary for the existence of any form of life and Darwinian evolution, in particular. The conclusions may also help to throw further light on other suggested theories and observations in physics and the role of quantum decoherence in life and its adaptive ability.
Keywords: causality, cosmological time, entropy, evolution, fluid dynamic modelling.

WIT Transactions on State of the Art in Science and Engineering, Vol 90, © 2016 WIT Press
www.witpress.com, ISSN 1755-8336 (on-line)
doi:10.2495/978-1-78466-155-7/001

1 Introduction

Recent advances in fundamental physics demonstrate that we live in a world of bewildering complexity. Whereas some topics illustrating this point, such as particle physics [1], are well covered, this is not the case in some other areas. Prominent among these is gravitation, a force field which cannot be shielded like electromagnetic fields, yet seems to emanate uniformly from all forms of matter as one of the smallest, but most important, of forces known to us. Our understanding of gravity has hardly advanced since Einstein's General Theory of Relativity.

The other great outstanding mystery in physics today is the nature of time itself, a subject which has intrigued us since the time of Aristotle [2]. Once again, our understanding of the nature of time is minimal despite the huge literature [3–7] devoted to it. This paper reports a relevant observation, which may lead to a better understanding of a subject that has puzzled us for centuries.

2 Fluid dynamic modelling

This well-established technique is covered in most text-books on fluid dynamics [8, 9], which include the use of Buckingham's 'π' theorem [10]. Many investigations of problems in science and technology require a partial modelling approach [11, 12]. However partial modelling presents obfuscation in clarifying the conclusions in this paper. This is because when too many dependent variables are involved, these often entail their own separate and yet constant interacting time scales which would confuse our interpretation. To explain clearly how fluid dynamic models can help to elucidate the nature of time, it is best to revert to the original study based on the experiments of Reynolds [13].

3 Smooth duct flow and the dimensionless Navier–Stokes equations

Incompressible fluid flow of a Newtonian liquid is expressed by the Navier–Stokes equations, shown here in vector form. The equations are non-linear.

$$\rho\left(\frac{\partial \mathbf{v}}{\partial t} + \mathbf{v}\cdot\nabla\mathbf{v}\right) = -\nabla p + \nabla\cdot\mathbf{T} + f \tag{1}$$

where **T** is the stress tensor of the effect from viscosity.

Here we are mainly interested in the direction of flow down a pipe in the 'i' direction:

$$\rho\left(\frac{\partial u_i}{\partial t} + u_j\frac{\partial u_i}{\partial x_i}\right) = -\frac{\partial p}{\partial x_i} + \mu\frac{\partial^2 u_i}{\partial x_j \partial x_j} + f_i \tag{2}$$

In dimensionless form, eqn (2) can be expressed with greater clarity [14]. The 'f' forces are zero because there is no free surface in the flow.

$$\frac{\partial u_i^*}{\partial t^*} + u_j^* \frac{\partial u_i^*}{\partial x_j^*} = -\frac{\partial p^*}{\partial x_i^*} + \frac{\upsilon}{UD}\frac{\partial^2 u_i^*}{\partial x_j^* \partial x_j^*} \tag{3}$$

In eqn (3), $u_i^* = u_i/U$; $u_j^* = u_j/U$; $t^* = tU/D$; $x_i^* = x_i/D$; $x_j^* = x_j/D$; $p^* = p/\rho U^2$; $\upsilon = \mu/\rho$; $UD/\upsilon = N_{Re}$.

It should be noted (Appendix I) that the term $\partial p^*/\partial x_i^*$ is a function of N_{Re} only. It should also be noted that because the same eqn (3) is used for all models and the equation has the same boundary conditions in non-dimensional form, then the solutions to eqn (3) are identical for any particular value of N_{Re}. This applies to the solution in closed form (laminar flow) and also in turbulent flow where u_i, u_j and u_k are all statistically varying functions of time as well as space. In turbulent flow, solutions to eqns (2) and (3) have to be obtained numerically.

Use of eqn (3) does not depend on the cross-sectional shape of the pipe, although our examination will be restricted to flow well down a circular pipe where the boundary layers [15], whether laminar or turbulent, have merged and become fully developed. In this way, eqn (3) does not need to contain an extra dimensionless parameter in terms of D/L_{total}, the presence of which, again, would confuse the analysis.

It is seen that in terms of the dimensionless parameters used in eqn (3), the flow detail in dimensionless form depends only on the Reynolds number. Nevertheless, this claim has a caveat.

In the range $0 \le N_{Re} \le 2000$ where the flow is laminar and where the flow is turbulent ($2800 \le N_{Re,}$), the conclusions arising from eqn (3) are valid. However, 2800 may become a higher value if steps are taken to delay the onset of turbulent instability. Where N_{Re} lies in the transition range [15, p. 526, Fig. 20.1] where the flow changes over from laminar to turbulent, extra parameters causing instability, enter eqn (3). Typically these are pipe imperfections like roughness or small vortices (fluid circulation) which enter the pipe. Today we would describe this instability as a form of 'butterfly effect', interpretable by 'chaos' theory, but one which is in a very severely constrained system. The transition range of N_{Re} brings in extra parameters such as pipe roughness effects into eqns (2) and (3) [8, p. 286, Fig. 5.32].

Now consider the flow of the same fluid in different-sized ducts but with the same Reynolds number. Because the kinematic viscosity (μ/ρ) is the same, then UD is the same in pipes with the same value of N_{Re}. If we consider two pipes denoted by suffices A and B, then:

$$U_A D_A = U_B D_B \tag{4}$$

Also, because eqn (3) is the same for pipes A and B and because the pipe flows have the same boundary conditions in dimensionless form, then the solutions to eqn (3) must be identical. This conclusion applies if the solutions are in closed form (laminar flow) or are statistical (turbulent flow where the flow is expressed in statistical parameters). It so happens that these dimensionless statistical values do not vary much when $N_{Re} \ge 5 \times 10^4$ [15, p. 563].

The important conclusion is that any one of a series of such identical models can be considered to operate in real time. It then follows that all the other models behave identically but function in different time scales, either faster or slower. For the simple case of a Newtonian fluid flowing in a circular duct, it follows that these time scales are related by a very simple expression:

$$T_A/T_B = (D_A/D_B)^2 \quad (5)$$

This expression applies to both laminar and turbulent flow.

Equation (5) and its derivation are significant. The derivation suggests that time is a flowing phenomenon, a conclusion which most of us would regard as mere common sense. However, it is quite common in modern physics for common sense and reality to fail to concur. Nevertheless, the observation that the passage of time can be varied by these models does not support the popular theory [16, 17] that time is a mere ingrained figment of the human imagination. Yet, the terms in eqn (5) are those which would be expected as they only relate to space-time. Thus, it seems that time can be interpreted and used both as a flowing phenomenon and also as a part of space-time. In physics, such dual conclusions are sometimes forced upon us. Wave–particle duality is probably the best known example which springs to minds.

3.1 Experimental confirmation

The theoretical approach described in Section 3 can be readily subjected to experimental confirmation. This is reasonably straightforward because this would be essentially just a much larger repeat of Osborne Reynold's original study, but using modern instrumentation. To be able to demonstrate, and particularly over a range of artificially induced time scales, that time is a constant, forward-moving variable, is an important contribution to science and technology despite the fact that this is a supposition we have been using for many years. The various forms of turbulent correlation coefficients, Reynolds stresses, etc. would be expected to demonstrate similarity in all the models examined of different sizes [18, pp. 39–67].

Flow of a fluid in a pipe is a convenient way of transforming time to the spatial dimension. The past, the present and the future are all different positions of distance down the pipe, expressible approximately in terms of U. The arrow of time is in the direction of fluid flow. For the limited conditions specified above, the flow in a pipe effectively freezes the past, present and future for observation into the same instant of time. For turbulent flow, where the increments in time are related statistically, this is an interesting facility in its own right.

Instrumentation techniques for studying turbulent flow have improved enormously in recent years and a laser-activated tracer technique [19] would be a vast improvement on early methods [20] used to avoid disturbing the flow from invasive instrumentation. Significant improvements in data analysis are also available in more recent work [21].

3.2 Time scales in other dimensionless numbers

Although it has been necessary to simplify the dimensional analysis in this paper to include N_{Re} as the only, and yet most relevant variable, usually most engineering problems use partial modelling where the other dimensionless parameters are also often (but not always) found to be different time scales associated with other physical details. Sometimes these significant dimensionless parameters are too numerous [11] for the dimensional analysis to be useful. At other times, the time scales are too difficult to define mathematically. One such example is an elastic tube (in this case, a human artery) which is tethered elastically to other human tissue and also subjected to pulsatile flow [12]. The Womersley number used only goes part of the way in dealing with this problem.

Time scales have been found useful in other areas of biology. An interesting example [22] is the claim that a time scale based on 'mean generation time of species' instead of 'years of life' reduces 'noise' in the statistical analysis of data related to endangered species.

Also, it should be noted that ref. [12] used the normal technique of evaluating the order of magnitude of different parameters in order to assess their likely significance. The present study presents an alternative way of evaluating the likely significance of parameters in dimensional analysis. This is to examine their time scales in the analysis.

As an example, consider chemical reprocessing of spent nuclear reactor fuel elements. The processed mixture contains a large range of radioactive isotopes which decay at widely varying rates depending on the quantum effects. It is likely to be instructive to compare their half-lives with the time scales of the chemical reactions involved. It may be possible to ignore isotopes with short half-lives and regard isotopes with long half-lives as a constant factor in design analysis. In this way, we may be able to simplify a very complicated engineering design realistically.

4 The qualitative nature of time

The above fluid dynamic approach expresses time scales as ratios in terms of other parameters, but this tells us nothing else about the present time in which we live. At this stage, we can only conclude that time, be it real, or even artificially engendered in the human mind, always appears to move at a constant rate, but it does not appear to be related to a fundamental constant like G or c.

Clearly we need to extend our investigation. Prigogine and Stengers [23] have drawn attention to the possibility of a relationship of time with the velocity of light given by Maxwell's relationship:

$$c = (\varepsilon_0/\mu_0)^{-1/2} \tag{6}$$

but this seems to have led nowhere.

In advancing our investigation, it is considered that our prime need is for a plausible physical model on which a detailed mathematical model and analysis can be based. Previous studies [23–26] indicate that we are well-equipped with regard to mathematical techniques. However, as shown below, this is not the case because we need to use non-equilibrium thermodynamic calculation techniques and this subject is still in its infancy [27]. The likely success of mathematical modelling is critically dependent on the choice of model we seek to analyse. Therefore, the rest of this paper will concern itself with this necessary preliminary stage of further research.

The role of the explorative experimental physicist is a vital precursor to any full analysis by mathematical physicists. This is needed in the present study as mentioned in section (7.1). Experimental facts need to be verified by laboratory physicists first and tested rigorously with an investigation of their likely range of application. Strange anomalies need to be reported, including suspicions concerning possible singularities and complex numbers which may have far-reaching effects in mathematical analysis.

The simplified preliminary discourse in Appendix II may help the readers in following the arguments to be suggested.

To our earthly experience, time can be divided into three parts: the past, the present and the future. From the point of view of the physicist, they seem to be three very different regimes.

4.1 The past

The main characteristic of the past is its permanence. If we were to keep comprehensive historical records, we would be able to analyse all past events in terms of classical Newtonian mechanics. Quantum effects are apparent, but they are frozen in time to make causality certain and examinable. Perhaps the most important and valuable quality of the past is that it appears to be dead and in a permanent state. It is fully deterministic. Its entropic state as a whole system is always less than (or at the most equal to) the entropy of the immediate present. Its entropic state is always less than that in the future.

The past is of enormous value to a cognizant species such as *Homo sapiens* in that it affords us the facility of experience to plan for events in an uncertain and indeterminate future. For other biological species, the past is potentially a detailed record of the mechanism of their automatic Darwinian adaptation to their own uncertain future.

4.2 The future

The future is so very different from the past. It is hardly ever deterministic, but is subject to widespread forms of statistical uncertainty. Disease, fickle climate and conflict are typical examples. To a small extent, uncertainty is limited when probability can be neglected for practical purposes. The behaviour of the solar system is one such example, although it also changes slowly. Most of our future experiences are statistically based. The future is also the true realm of quantum physics, natural biological control systems and uncertainty. We can make plans for our own future, but only with detailed analysis of the random nature of natural processes. The future is always associated with increase of the entropic state of the whole system.

4.3 The present

The present, or instantaneous time, divides the past from the future. Because the past and the future are so very different, consideration of the present instant suggests to us to that the present time, although very short in duration, must have some very significant and unusual qualities which may not be obviously apparent to us. We need to ask ourselves what are the physical processes at work in present time? Using our own concept of reality, the present appears to move with our own existence continuously and at a constant rate. It appears to separate the past from the present and sweep all that still exists into an uncertain future. Although it may seem intangible to us, instantaneous time suggests to us that it is likely to be a barrier between the past and the very different future. Physically it is certainly a very pronounced discontinuity. Viewed very simplistically, the present seems to be a moving temporal barrier between the past and the future and one where all processes, be they physical, chemical or biological, change dramatically. Perhaps this barrier may become expressible in units of Planck time?

$$t_p = (hG/c^5)^{1/2} = 5.39\ldots \times 10^{-44}\ \text{s} \tag{7}$$

The present is where the frozen immobility of the dead past suddenly presents itself as still alive and becomes subject to statistical mechanics and the second law of thermodynamics involving increasing entropy and uncertainty. It is this natural moving coordinate system which needs close theoretical scrutiny.

5 Importance of Prigogine's work

The above reasoning that instantaneous time is likely to be an entropic barrier separating past time from future time is based largely on the selected opinions of several, but not all, scientists. Prigogine and Stengers [23] have used very different arguments to come to the same conclusion [23, pp. 295–296] that an entropic barrier is an important aspect of instantaneous time in its presumed function to separate the past from the future. Prigogine's research has received considerable independent support [28]. This agreement is encouraging.

Any attempt at theoretical examination of the future does not seem to be promising because, by its nature, the future cannot yield firm data. On the other hand, comparing the past with the present may help in our understanding. In examining the past, it seems that the 'sliced loaf of bread model' [29, pp. 57–61, pp. 130–139], which is a development of the 'block' model, seems to be particularly useful in investigating an entropic time barrier model.

One likely consequence of this model is that the entropic transfer barrier from the present to the past seems to be formidably strong compared with the similar entropic barrier if moving from the past to the present. It can be likened to a non-return valve except that it is one controlled by entropy (and/or some other factor such as possibly a quantum decoherence effect – see section 7.1). The forward barrier exists, but transfer across this barrier is from a low entropy state to a higher one, which is easier thermodynamically This barrier is especially weak for an inanimate object which, in an isothermal transfer, requires almost no entropy change at all in making a transfer, but the barrier is much stronger for a living object which needs to excrete entropy to the environment during any suggested possible transfer. Thus, the possibility of transfer from the present to the past appears to be by far the most difficult effect to achieve. This may be because each increment (obliquely sliced segment of the sliced loaf of space-time [29, p. 135]) of past time isolates the passing increment of time from the adjacent increment which succeeds it in forward-moving time. This barrier can, perhaps, be considered to be a multi-layered one, thus making it likely to be a very strong one indeed.

In conclusion, it seems likely that any living creature (and this includes even a well-designed automated machine – which also would have entropy dispersal needs) cannot pass from the present to the past. It seems that this may explain away the paradox of a person returning to the past to kill his/her own grandfather. It seems that even a very cleverly designed machine could only have a very low penetration into the past. It may be reassuring for us that our ancestors appear to be safe in their graves. Nevertheless, it is necessary to carry out calculations to clarify this supposition.

Perhaps the strongest evidence for the strength of this backward moving part of the entropic time barrier is the oft-quoted question, 'Where are the tourists from the future?'

When we consider the possibility of travelling from the present to the future, a similar entropic barrier exists, although it seems to be less of a hurdle as it is from a lower entropy state to a higher one. However, it has been pointed out that such transfers are not just a matter of net entropy transfers, but the rate at which entropy can be transferred becomes important [30]. This is why we cannot avoid using the latest developments [27] in non-equilibrium thermodynamics when making calculations. Thus, the possibility of such a transfer may be limited by the size of the time barrier, if it can be shown to exist.

It seems that although time travel to the future may be difficult, it cannot be ruled out as a possibility. However, it has been pointed out [3] that the future may not contain any events for us to visit. The main problem, however, seems to be that time travel to the future seems likely to be a one-way trip because of the seemingly much stronger barrier preventing return.

Physicists remain perturbed [29, pp. 128–142, 31] by the fact that the behaviour of time is almost unique among our physical laws in not being symmetrical about itself (time reversal invariant). It has even been suggested that if time can be reversed, then entropy should increase as time moves backwards [29, p. 161, Fig. 6.2b]. This has been questioned [32].

It seems possible that the natural symmetry of time may have changed at some stage in early creation towards a forward-moving arrow of time. However, opinions vary in such a difficult speculation [33]. Our world has endured many strange events before it became fit for intelligent life-forms to develop. Sometimes we have plausible explanations such as the necessary demise of the dinosaurs to make way for the development of mammalian life. Usually reasons for past events remain shrouded in mystery.

6 Present controversy relating to our interpretation of the nature of time

The observations in section (3) suggest that time can be interpreted either as space-time as elucidated by Einstein's General Theory of Relativity (this is the Parmenidean School) or alternatively, time can be interpreted as a flowing phenomenon. These are the Heracliteans [3]. Normally different people subscribe to one or the other of these viewpoints and only a minority seem to subscribe to both.

Design engineers are always 'Heracliteans' [3], although few professional engineers would have ever heard of the term. Design engineers use a whole range of equations expressed in terms of time as a constantly flowing continuous parameter (the 'time flux' point of view). They integrate and differentiate these expressions with success in different branches of technology. Often these design equations are of an empirical nature with limited scientific basis. It is sufficient for these Heracliteans that their designs can be brought to a satisfactory and logical conclusion.

On the other hand, the majority of physicists seem to be 'Parmenideans' [3]. Only a few [3, 34] declare themselves openly to have Heraclitean sympathies. Parmenideans assert that time does not flow at all, and some even suggest that time does not even exist and is a mere figment of human imagination [16, 17]. Many Parmenideans are eminent scientists [29, 35] and include Einstein himself. They support the 'stasis' view of the 'block universe'. The pedigree of the Parmenideans is beyond reproach because their viewpoint is to examine all existence from outside of time and space. This is the presumed divine creator viewpoint, expressed so lucidly by C.S. Lewis [36]. This metaphysical viewpoint has a history many centuries old including contributions from theology scholars and other distinguished philosophers.

It seems reasonable to conclude that a deeper understanding of the nature of time is needed.

7 Suggested outline of further investigation

Statistical mechanics and non-equilibrium thermodynamics [27], in particular, can be used to calculate the desirable width of an entropic barrier in terms of Planck time. This would be on a probability basis, i.e. the chance that instantaneous time can form an effective barrier to a prescribed transfer process. Calculations would involve different materials and different particles. It is not sufficient to merely examine the transfer of particles. It is also necessary to evaluate their ability to transfer information. Information [37, 38] has been described as negative entropy. Fundamental particles can pop into and out of existence all the time, but their ability to carry information is limited. This suggests that the required width of the entropic barrier could be best evaluated for a small virus with an estimation of the probability of its survival in an entropic transfer. This may be of basic interest because a virus is a basic form of information. Can a virus cross an entropic time barrier from the present to the past and vice versa? This can be compared with the smaller width barrier needed to block transfer likely to be related to more sophisticated life-forms. These calculated widths would be of much interest compared with the statistical consequences of a different entropic width which might actually exist. In other words, is a natural entropic width compatible with the needs of our earthly existence? Would this throw light on the time scale of the natural flow of time which we experience? What is the actual probability of intact information transfer for different conditions? Would this information be related to other – suspicions suggesting that the exact requirements for a 'platform for life' are being met [39–41]? Other suspicions that Prigogine's entropic barrier may be more than a discontinuity and may exhibit singularity properties and complex numbers arising from quantum considerations are also worth bearing in mind.

7.1 The need for prior experimental investigation before investigating Prigogine's entropic barrier model

The present writer is unaware of any serious laboratory investigation into the physical properties of instantaneous time. This may be for a very good reason that the experimental difficulties are too great. Research into the allied field of quantum tunneling [42] suggests that this is the case. Quantum tunnelling is very fast, apparently exceeding the speed of light (although careful examination has shown that this does not violate relativity and causality considerations).

Prigogine's approach may be the best candidate for finding explanations, but there may be other explanations of the strange physical nature of instantaneous (present) time. These may be of a quantum mechanical nature based on Schrodinger's wave equation and decoherence effects [43]. Before embarking on an expensive investigation into suspected entropic time barriers, it seems to be wise to carry out some laboratory studies first. We have a range of equipment and techniques for measurements over short intervals of time, etc. to help us to investigate the minimum time interval needed for an entropic time barrier to explain the transition of present time to past time. If there is no experimental evidence at all for finite time intervals in the nature of instantaneous time, then it would be preferable to delay an expensive numerical investigation into entropic time barriers in favour of finding an alternative quantum explanation. Quantum effects are expected to act almost instantaneously and much faster than an entropic effect. It is difficult to restrain quantum effects from collapse and an event lasting 10^{-4} s is difficult to achieve [44]. Other fundamental studies in solid state physics may also be useful in this investigation.

At this early stage, it is also important to consider that decoherence-controlled and entropy-controlled transfer processes may even act together jointly in giving us the mechanism of a suspected barrier in time. When we know so little, it seems wise to keep our options open.

Many quantum processes are time-reversible. For example, a container of inert gas like neon can, theoretically, be kept for all eternity at is maximum steady entropy but still experiencing statistically varying conditions, as the molecules collide elastically with each other. Yet, quantum physics also has some very irreversible phenomena such as radioactivity and wave collapse/decoherence. It is surmised that this complexity may be an important ingredient in our 'platform for life'.

Furthering our understanding of the physical nature of time would be an important contribution towards our understanding the 'platform for life' on which we all depend. Our very existence seems to be the result of a long sequence of statistically improbable freak events [39–41]. If we are eventually forced into accepting teleological explanations for our existence, we should at least have tried hard first to elucidate and understand the scientific basis for our 'platform for life'.

7.2 The need for plausibility in developing a theoretical model

Much of the research in modern physics involves phenomena (and even presumptions) which we cannot see. For example, only recently has it become possible to see the statistical variation of the orbiting electron in a hydrogen atom [45]. In another example, Schrodinger's wave function appears to be no more than a mathematical abstraction, and yet its application generates realistic results which can be confirmed experimentally.

What is needed for elucidating Prigogine's entropic barrier is a plausible and well-founded theoretical basis. It is likely that extensive numerical methods will be needed to examine such a theory. Altering parameters in computer calculations can throw considerable light on the accuracy of any theory and may even improve our ability to develop the theory. But this is only possible if the basic theory is plausible in the first place and includes all relevant parameters. Repeated calculations will sometimes close such a study to formulate reliable equations. These methods are lengthy and expensive as mentioned in an earlier recent paper [46, pp. 451–453] regarding optimisation of a new technology, which we will probably be forced to develop as a consequence of adverse influence on world climate.

8 Discussion

Looking objectively at our own existence, we find that we are compelled to live lives which are very severely constrained. Our best form of communication (by electromagnetic waves) is tediously slow at 3×10^8 m sec^{-1}. It limits us to viewing much of the history of the cosmos which is only long-past. At its present rate of universe expansion, it seems that we are destined to even lose this part of history. We might well look with envy at quantum-entangled particles and their apparently instantaneous communication. In another example of our limitations, almost all of the entire chemistry of our bodies is controlled only by the selective power of outer electrons in the atoms we have inside us. Yet, perhaps, the greatest constraint in our lives is the second law of thermodynamics which chains us to entropy and the forward arrow of time.

Yet, recognising our own humble status as mere integrated examples of organic chemistry, perhaps we should be grateful for our stable and secure 'platform for life'. Our 'platform for life' has its set of scientific rules resulting in causality, so that we can be logical in explaining our experience and managing our affairs. Most of these rules are workable rules in classical Newtonian mechanics. For example, these rules are sufficient for us to engineer an automated exploration of the planets in our solar system which is an impressive achievement. Yet, it is mainly the field of quantum physics which shows us that rules often have exceptions and often these exceptions seem to be necessary.

9 Conclusions

It is concluded that when comparing some similar systems, time can be shown to move faster or slower than the time scale in which we live. Although, like c, time is a constant in our lives, it seems that it is not related to a fundamental constant like G, the gravitational constant.

Studying fully developed turbulent flow transforms time events of past, present and future into sequentially identifiable positions down a long tube. At any particular instant, the past, present and future exist together and are related with each other statistically, but each observation at any instant can be engineered to a time scale to suit us.

Prigogine's earlier research seems to be a promising line of investigation to take in explaining the true nature of time by evaluating the influence of entropy as an explanation of the strange time asymmetrical phenomena which we observe. It is also reasoned that time, at the present instant, presents a strong discontinuity between the past and the future because of the very different physical nature of time-past and time-future. By way of explanation, there are also suspicions that a quantum decohesion effect, which we do not yet understand, may also be involved in the discontinuity effect in instantaneous time.

Clearly, this topic and its consequences deserve a more detailed examination than can be given here. It is proposed to review the subject again in 5 years [47]. Five years may seem a long time in a fast-moving scientific era, but it is short in comparison with several hundred years of mystery attached to the subject of time.

It seems that time can be examined realistically in two ways. Time is examinable as a part of space-time, which is necessary if relativistic effects are felt. Alternatively, in our everyday world, the concept of time as a moving parameter helps us to understand and manipulate our world. The conclusions resulting from section (3) support this. Both the Parmenidean and Heraclistic viewpoints are useful.

As a final conclusion, the forward-flowing nature of time is clearly a most important factor in understanding the mechanisms at work in our 'platform for life'. This platform is so intricate, complicated, elegant and sophisticated that it is hard to avoid the suspicion that it is also contrived.

Appendix I: $\partial p^*/\partial x_i^*$ is a function of N_{Re} only

Consider a length x_i of circular pipe with diameter D. This incurs a pressure drop $p(x_i) - p(0)$, where $p(0)$ is the pressure where $x_i = 0$. In this region, the boundary layers are fully developed, and therefore, $p(x_i)$ varies linearly with x_i. Also, τ, the wall shear stress on the fluid, is constant.

Balancing the forces on a moving cylindrical element of the fluid (which is not accelerating), we have

$$\pi D^2 (p(x_i) - p(0))/4 = \pi D x_i \tau \tag{8}$$

where $\partial p(x_i)/\partial x_i = 4\tau/D$.

The wall shear stress (τ) for a smooth pipe is given by

$$\tau/\rho U^2 = \Phi(N_{Re}) = f, \tag{9}$$

the friction factor [8, p. 288].

Thus, $\partial p/\partial x_i = 4\rho U^2\Phi(N_{Re})$ and

$$\partial p^*/\partial x_i^* = 4\Phi(N_{Re}), \tag{10}$$

i.e. a function of N_{Re} only.

Note also that U is normally taken as the mean axial velocity in the pipe evaluated from integration of the radially varying velocity profile in the direction of the pipe axis.

For laminar flow, $u_i(r)$ is parabolic and for turbulent flow, the time mean average of $u_i(r)$ is taken from experimental data. Any other consistently used and reasonable definition of U is suitable.

Appendix II: simplified outline of a proposed basis for physical/mathematical analysis into instantaneous time and past time

The natural coordinate system for any mathematical analysis of time seems to be a moving Lagrangian coordinate system which follows us exactly. Instantaneous time (now) also seems to be bounded by two barriers.

- Future time which is subject to all the vagaries of statistics (and almost all of our very existence is statistically based). Future time seems, therefore, to be the most unattractive as a direct basis for analysis.
- By contrast, the supposed barrier between instantaneous time and past time has hard and determinate facts against which we can test any theory. Therefore, this seems to be the logical path to take in research.
- The approach is best illustrated by the following story about 'grandma and her chair' and how this is affected by the lower entropy state of the immediate past and the higher entropy state of the present time.

Grandma's chair has no difficulty in travelling intact from the past to the present. The chair is inanimate and large. The chair's entropy has hardly changed at all in passing from the past to the present. Quantum effects are immeasurably small when considering large inanimate objects. How does this compare with Grandma herself when she died? Certainly, Grandma's body had no difficulty moving from the past to the present. Nevertheless, its entropy increased considerably in moving from the past to the present. So, for a different reason, her body had no difficulty in moving from a lower entropy past to a higher entropy present. But what about the life of Grandma which was clearly debarred from entering the present? Throughout her life, Grandma kept her entropy more or less constant by increasing the entropy of her immediate environment. As she got older, this ability declined. She probably died when her immune system faltered and with this, her ability to control her entropy declined.

However, all this speculation is very complex, so let us consider a very small piece of matter which is easier to examine mathematically.

Many small pieces of matter are in a state of thermodynamic equilibrium where entropy is at its maximum possible value. Such an element would have no trouble moving from the past to the present (which is at a higher general entropy). In this case, no significant barrier seems to exist between the immediate past and the present.

It seems that for small objects not in a state of maximum possible entropy, the probability of moving from the past to the present or vice versa can be evaluated mathematically using statistical mechanics. Thus, if entropy conditions are sufficiently favourable, crossing the presumed time barrier may be possible.

It seems that the 'temporal width' of this barrier can be estimated (perhaps in units of Planck time). This temporal width is clearly a variable depending on the physical properties of the particle of matter being considered and its surroundings. It seems that life-forms, even very small viruses, sometimes cannot pass this barrier except with a very small probability.

At this stage we can suspect that, although the barrier between immediate past and present in invisible, it exists and it can be examined mathematically using statistical mechanics. The barrier seems to depend on the object considered and whether the object is inanimate or living. The supposed temporal width of the barrier is an important parameter because this affects the probability of passing between past and present and vice versa, and this probability seems likely to decrease as the past recedes in real time from the present.

Nomenclature

c = velocity of light (eqn (6) where ε_0 and μ_0 are the permittivity and permeability of free space, respectively)
f = friction factor $t/\rho U^2$. N.B.: in eqns (2) and (3), f refers to a non-existent force
L_{total} = distance down the pipe where boundary layers have merged to give uniform flow
N_{Re} = pipe Reynolds number UD/υ
p = pressure in pipe
t, t_p = real and Planck time
T_A/T_B = relative time scale ratio in pipes A and B
u.v, = fluid velocities in pipe
U = average axial flow velocity
x = axial position in pipe

D = circular pipe diameter ($2r_{max} = D$)
ρ = density of fluid
$\upsilon = \mu / \rho$
μ = viscosity
t = wall shear stress of pipe on fluid (negative in coordinate system used)
F = functional relationship

References

[1] Martin, B.R., *Particle Physics,* Oneworld: Oxford, pp. 1–201, 2011.
[2] Davies, P.C.W., *About time – Einstein's unfinished revolution,* Penguin: London, p. 29, 1995.
[3] Weinert, F., *The march of time-changing conceptions of time in the light of scientific discoveries,* Springer: Dordrecht, pp. 1–281, 2013.
[4] Hawking, S. & Penrose, R., *The nature of space and time,* Princeton University Press: Princeton, pp. 1–156, 2010.
[5] Zeh, H.D., *The physical basis of the direction of time (Edn 5),* Springer: Berlin & Dordrecht, pp. 1–240, 2007.
[6] Reichenbach, H., *Direction of time,* University of California Press, pp. 1–280, 1956.
[7] Gibbons, G.W., The emergent nature of time and the complex numbers in quantum cosmology, In Mersini-Houghton, L. & Vaas, R.,(eds) *The arrows of time: a debate in cosmology,* pp. 109–145, Springer: Dordrecht, 2012.
[8] Streeter, V.L. *Fluid mechanics, 5th Edn,* McGraw Hill: New York, pp. 201–231, p. 286, p. 288, 1971.
[9] Kay, J.M. & Nedderman, R.M., *An introduction to fluid mechanics and heat transfer, 3rd Edn,* Cambridge University Press, pp. 56–71, 1974.
[10] Buckingham, E., Model experiments and the form of empirical equations, *Trans ASME,* **37**, pp. 263–296, 1915.
[11] Boothroyd, R.G., Similarity in gas-borne flowing particulate suspensions, *Trans. ASME (Amer. Society of Mechanical Engineers (Journal of Engineering for Industry)),* **91(2)**, **series B**, pp. 303–314, May, 1969.
[12] Hutton, P., Foster, K., Ashton, F. & Boothroyd, R.G, An appraisal of a modelling technique for arterial flows, *Engineering in Medicine,* **4(3)**, pp. 3–12, 1976.
[13] Reynolds, O., An experimental investigation of the circumstances which determine whether the motion of water shall be direct or sinuous and the laws of resistance in parallel channels, *Trans. Roy Soc. London,* **174**, pp. 935–982, 1883.
[14] Kopera, M., Non dimensionalisation of the Navier–Stokes equations, Centre for Scientific Computing, University of Warwick www2.warwick.ac.uk/fac/sci/eng/study/…/ non-dimentionalization.pdf
[15] Schlichting, H., *Boundary layer theory,* McGraw Hill, pp. 1–747, 1968.

[16] Barbour, J., *The end of time: the next revolution in our understanding of the universe,* Weidenfeld & Nicholson: London, pp. 1–371, 1999.
[17] Slezak, M., The time delusion, *New Scientist,* **219**(#2937), pp. 34–38, 2nd November. 2013.
[18] Hinze, J.O., *Turbulence an introduction to its mechanism and theory,* McGraw Hill: New York, pp. 39–67, 1959.
[19] Tropea, C., Yarin, A.L. & Foss, J.F., (eds) *Springer handbook of experimental fluid mechanics (vol. 1)* Springer: Dordrecht, pp. 366–368, 2007.
[20] Boothroyd, R.G., An anemometric isokinetic sampling probe for aerosols, *J. Sci Instruments,* **44**, pp. 249–253, 1967.
[21] Choi, J-I, Yeo, K, & Lee, C., Lagrangian statistics in turbulent channel flow, *Physics of Fluids,* **16(3)**, pp. 779–794, 2004.
[22] Frankham, R. & Brook, B.W., The importance of time scale in conservation biology and ecology, *Ann. Zool. Fennici* **41**, pp. 459–463, 2004.
[23] Prigogine, I. & Stengers, I., *Order out of chaos: man's dialogue with nature,* Bantam: Toronto, pp. 257–296, 1984.
[24] Prigogine, I., *From being to becoming: time and complexity in the physical sciences,* Freeman: San Francisco, pp. 1–272, 1980.
[25] Prigogine, I., *Introduction to thermodynamics of irreversible processes (2nd Edn)* Interscience: New York, pp. 1–119, 1961.
[26] Prigogine, I., *Non equilibrium statistical mechanics.* Interscience: New York, pp. 1–319, 1962.
[27] Attard, P., The second entropy: a general theory for non-equilibrium thermodynamics and statistical mechanics, doi 10.1039/b802697c, *Ann. Rep. Prog. Chem. (C),* **105**, pp. 63–173, 2009.
[28] Coveney, P. & Highfield, R., *Arrow of time: a voyage through science to solve time's greatest mystery,* Allen: London, pp. 161–378, 1990.
[29] Greene, B., *The fabric of the Cosmos,* Penguin: London, pp. 1–569, 2004.
[30] Gold, T., & Bondi, H., (eds) *The nature of time,* (mtg May–June, 1963) Cornell University Press: New York, p. 229, 1967.
[31] Lederman, L.M. & Hill, C.T., *Symmetry and the beautiful universe,* Prometheus: New York, pp. 1–363, 2004.
[32] Penrose, R., *The road to reality: a complete guide to the laws of the Universe,* Vintage books: London, p. 698, 2004.
[33] Harland, D.M., *The big bang: a view from the 21st century,* Springer Praxis: Chichester, UK, p. 233, 2003.
[34] Smolin, L., *Time reborn: from the crisis of physics to the future of the universe,* Allen Lane: London, pp. 1–305, 2013.
[35] Davies, P.C.W.D., Stirring up trouble, in Halliwell. J.J., Perez-Mercader, J. & Zurek, W.H., (eds) *Physical origins of time asymmetry,* Cambridge University Press, pp. 119–130, 1994.
[36] Lewis, C.S., *Mere Christianity,* Collins: London, pp. 145, 1944.
[37] Jarzynski, C., Equations and inequalities: irreversibility and the second law of thermodynamics at the nanoscale, *In Duplentier, B., (ed) Progress*

in mathematical physics (vol. 63) (15th Poincare' seminar, 2010 Springer: Basel, pp. 145–168, 2013.

[38] Fromhold, A.T., Jr, *Quantum mechanics for applied physics and engineering*, Academic Press: New York, pp. 237–278, 2012.

[39] Wilczek, F., *The lightness of being: big questions, real answers,* Penguin, London, pp. 1–270, 2008.

[40] Rees. M.R., *Just six numbers,* Weidenfeld &Nicholson: London, pp. 1–194, 1999.

[41] Denton, M.J., *Nature's destiny,* Simon & Schuster: New York, pp. 1–454, 1998.

[42] Prutchi, D., *Do it yourself quantum physics: exploring the history, theory and applications of quantum physics through hands-on projects,* J Wiley: Hoboken, pp. 178–180, 2012.

[43] Duplantier, B., Raimond, J-M. & Rivasseur, V., *Quantum decoherence Poincare Seminar*, (Progress in mathematical physics V 48), Springer: Birkhauser Verlag and Dordrecht, pp. 1–201, 2006.

[44] Al Khalili, J., *Quantum: a guide for the perplexed,* Orion: London, p. 249, 2014.

[45] Anon. Smile hydrogen atom you're on quantum camera, *New Scientist,* **218**(#2919), p. 13, June 1st, 2013. (Stodolna *et al. Physical Review Letters, doi.org/mmz).*

[46] Boothroyd, R.G. Proposed Australian transition to an anhydrous ammonia fuel transport economy to replace liquid petroleum fuel, *WIT trans. Ecology and the environment,* **186**, pp. 443–456, 2014.

[47] Boothroyd, R.G., On the physical nature of instantaneous and historical time: a review, boothroyd.org, 2020.

On computing the behavior of the mind from an eastern philosophical perspective

A. S. Karunananda[1], P. R. Goldin[2], G. Rzevski[3], S. Fernando[4] & H. R. Fernando[4]
[1]*General Sir John Kotelawala Defence University, Sri Lanka*
[2]*University of California, Davis, USA*
[3]*Centre for Complexity Science Applications, The Open University, UK*
[4]*Department of Computational Mathematics, University of Moratuwa, Sri Lanka*

Abstract

The human mind is one of the most intriguing complex systems found in nature. Research has been conducted to develop a computer model of the mind from the Theravada Buddhist description detailed in the *Abhidhamma* literature. Within this context, the mind is identified as a conditional flow of thoughts that is characterized by 52 mental factors, such as perception, volition, attention, anger, attachments, mindfulness and so on. Thus, mental factors are the building blocks of thoughts. Combinations of these 52 mental factors can form 89 different thoughts with characteristics such as feelings, roots, functions, doors, objects and bases. We have developed a computational ontology of Buddhist theory of mind (OntoBM) that describes the complex association among thoughts and mental factors. OntoBM can be primarily used to examine combinations of mental factors that can form thoughts, associations among mental factors and details of thoughts. More importantly, researchers and scholars can also use OntoBM to experiment with the effect of modulation of mental factors on the formation of various thoughts.
Keywords: artificial intelligence, Buddhism, complex systems, simulations.

1 Introduction

Nature is full of complex systems [1]. The brain and mind are excellent examples of complex systems. Although brain and mind are distinct terms, they have been

WIT Transactions on State of the Art in Science and Engineering, Vol 90, © 2016 WIT Press
www.witpress.com, ISSN 1755-8336 (on-line)
doi:10.2495/978-1-78466-155-7/002

used as synonyms in a wider spectrum of subject areas. Numerous brain simulators have been developed to carry out experiments that cannot be done otherwise with actual tissues and cells in a brain [2]. New research directions such as Mind uploading [3] and Blue Brain [4] are excellent attempts to model the brain. Numerous software programs [5] are also available to experiment with brain simulations. On the other hand, advancements in computing, neuroscience and psychology enable study of brain behavior through various technologies such as functional magnetic resonance imaging (fMRI) [6], electroencephalography (EEG) wave signals [7] and transcranial direct current stimulation (tDCS) [8]. Such studies have made a tremendous impact on understanding the structure and function of the human brain. However, limited research has been conducted on computer modeling and simulation of mind [9, 10]. In particular, very few studies have attempted to use [11, 12] computer modeling of mind from the Buddhist philosophical perspectives. We have conducted research to use *Abhidhamma* [13] in Theravada Buddhist theory of mind for the development of a computational ontology of mind. This paper presents the design and implementation of computational Ontology of Buddhist theory of Mind (**OntoBM**), which becomes the basis for developing a smart simulator of mind.

The rest of the paper is organized as follows. Section 2 reviews complexity science and computer modeling of complex systems. Section 3 discusses the Buddhist theory of mind. Section 4 presents the design of a computational ontology of mind. Section 5 presents how the ontology of mind can be used in practice. Section 6 concludes the paper with a note on further work.

2 What are complex systems?

Complexity science emerged as an attempt to explain and model what is called complex systems. This field has emerged from the pioneering work by Prigogine [14, 15]. Complex systems have been defined as the systems with large number of distributed and interconnected entities that manifest under lots of uncertainty in a dynamically changing environment [1]. The brain is a commonly cited example for a complex system, which comprises of 100 billion neurons [16] or processes that are massively interconnected, distributed and are constantly firing under dynamically changing chemical and physical reactions. The Internet, especially the Internet of Things, is also an excellent example for a complex system. As estimated, by 2015, there will be 25 billion devices connected to the -Internet of Things and billions of transactions move on the network of networks [17]. Apart from the brain, many other systems including industrial automation, aircraft maintenance and military logistic management are well-known examples for complex systems.

2.1 Computer modeling of complex systems

This section discusses suitable computational technologies for modeling complex systems. In this regard, we emphasize the importance of ontology and multi-agent technology.

2.1.1 Ontology

Traditionally ontology is the philosophical study of themes, such as being, existence and categories of being [18]. The computing community has borrowed the term ontology from philosophy. There are many definitions about ontology; yet in operational terms, ontology can be seen as a specific conceptualization of domain knowledge [18]. There are ontological solutions for various domains including medicine [19] and automotive industry [20]. There are also numerous tools available for ontology development, mapping, editing, merging and so forth. Protégé (generic framework for ontology development) [21], PROMPT (ontology mapping environment) [22] and automated solution for matching of two ontologies [23] are examples of such tools. It should be noted that ontology is a huge reservoir of knowledge or knowledge base. Thus, reasoning on ontology of complex domains gives rise to another challenge. There are also many tools (e.g. Chainsaw [24], FaCT++ [25] and Pellet [26]) for reasoning on ontology.

2.1.2 Multi-agent systems

Multi-agent systems [27, 28] technology can be seen as the best computational approach to model complex systems. In this context, all the entities in a complex system can be defined as small programs called agents. Each agent accomplishes its own goal, preference, knowledge and strategies to achieve a certain goal. Numerous examples can be given to show the power of multi-agent systems to model complex systems [1]. For instance, Hongqiao *et al.* [29] have used multi-agent systems for modeling complex hospital environments. Belgacem and Fates have presented a multi-agent solution for complex termite behavior [30]. Rzevski [28] has shown the application of multi-agent systems technology to model a variety of real-world problems including large-scale societies and huge technological and engineering systems. Numerous toolkits are available for the design and development of multi-agent systems. Probably, Java Agent Development Framework (JADE) [31] is the most popular among them.

3 Buddhist theory of mind

Buddhism, more specifically, Buddhist philosophy is primarily aimed at the development of mind in order to achieve freedom from suffering. The teaching of Buddhism comprises three baskets or sources of knowledge known as *Suttapitaka, Vinayapitaka* and *Abhidhammapitaka* [32, 33]. *Abhidhamma* is the most philosophical, analytical and structured presentation of the doctrine of Buddhism [32]. *Abhidhamma* is also regarded as Buddhist psychology [32, 34]. In literature, *Abhidhamma* comes under both Theravada and Mahayana traditions of Buddhism. Our research is based on the Theravada tradition of *Abhidhamma* which is rooted in the ancient Buddhist text known as *Abhidammaththa Sangah* [33]. *Abhidhamma* expounds a comprehensive description of 'mind' by showing definition, constitution, roles, mechanism, control and development of mind to achieve various cognitive skills.

3.1 Mind as a flow of thoughts

According to *Abhidhamma,* mind (*citta* in Pali) is a flow of thoughts. These thoughts are formed on the basis of various perceptions made through sight, sound, smell, touch, taste, thinking and recalling. Thus, in a broader sense, thoughts are of two types, namely, thoughts based on five senses and thoughts based on mind [13]. Each thought is conditioned by previous thoughts. Mind does not have a moment without a thought. To explain this phenomenon more specifically, Buddhist psychology has also identified a special type of thought, namely, life continuum (*bhavangacitta*) [13], which arises whenever a person does not manifest thoughts through five senses or the mind. For instance, when an individual falls into a dreamless sleep or becomes unconscious, the *bhavangacitta* manifests. The *bhavangacitta* is similar to a vehicle resting in a neutral gear when the engine is running. The Buddhist theory of mind suggests that the flow of thoughts in the mind occurs in a sequence of blocks of thoughts, which are known as *cittavithi* or thought processes. Each thought process begins and ends with *bhavangacitta.* The concept of thought process elegantly describes the mechanism of the mind. It is beyond the scope of this paper to discuss thought process in detail. A reader interested in thought process may refer to computer modeling of thought process by Karunananda [12].

3.2 Mental factors

Abhidhamma describes mental factors that in combination constitute different types of thoughts that arise in the mind [33]. There are 52 mental factors that determine the nature of thoughts. There are altogether 89 (or 121 according to a different classification) possible combinations of these 52 mental factors. As shown in Table 1, 52 mental factors are subdivided under three main categories: 13 ethically variables, 14 unwholesome and 25 wholesome. English translations for 52 mental factors vary. Thus, we have predominantly used the English translations given by Bodhi [33] in our research work.

As shown in Table 1, within 'ethically variables', there are seven universal mental factors that appear in all thoughts. These mental factors include contact (*passa*), feeling (*vedana*), perception (*sanna*), volition (*cetana*), one-pointedness (*ekaggatha*), psychic life (*jivithedriya*) and attention (*manasikara*). In contrast, 'occasional mental factors' appear in both wholesome and unwholesome thoughts. They are identified as initial application (*vithakka*), sustained application (*vichara*), decision (*adhimokkha)*, energy (*viriya*), zest (*priti*) and desire (*chanda*). Fourteen unwholesome universal mental factors exclusively appear in unwholesome thoughts. There are 10 more 'unwholesome occasional' mental factors that appear in certain thoughts, depending on whether the thought is related to greed, hatred and delusion. There are 25 mental factors which exclusively appear in wholesome thoughts. Within this category, abstinences, the illimitable and non-delusion are more related to higher-order thoughts pertaining to understanding of the ultimate truth.

Table 1: Fifty-two mental factors.

Ethically variables (13)	Unwholesome (14)	Wholesome (25)
Universals (7)	***Unwholesome universal (4)***	***Universal beautiful (19)***
Contact	Delusion	Confidence
Feeling	Shamelessness	Mindfulness
Perception	Fearlessness of wrong	Shame
Volition	Restlessness	Dread
One-pointedness		Non-greed
Psychic life	***Unwholesome occasional (10)***	Non-hatred
Attention	Greed	Equanimity
	Wrong view	Tranquility of mental states
Occasional (6)	Conceit	Tranquility of mind
Initial application	Hatred	Lightness of mental states
Sustained application	Jealousy	Lightness of mind
Decision	Avarice	Pliancy of mental states
Energy	Worry	Pliancy of mind
Zest	Sloth	Adaptability of mental states
Desire	Torpor	Adaptability of mind
	Doubt	Proficiency of metal states
		Proficiency of mind
		Rectitude of mental states
		Rectitude of mind
		Abstinences (3)
		Right speech
		Right actions
		Right livelihood
		Illimitable (2)
		Compassion
		Appreciative joy
		Non-delusion (1)
		Wisdom faculty

Any thought that arises in the mind can be characterized by a combination of these mental factors. However, some combinations of mental factors do not form thoughts. For example, attachment and hatred do not co-exist in a thought.

3.3 Thoughts

As stated earlier, thoughts are formed by mental factors, and flow of thoughts manifest the mind. However, the story about thoughts goes beyond these fundamental definitions of mental factors and thoughts. More specifically, *Abhidhamma* is not limited to describe the relationship between thoughts and mental factors. Thoughts are also described in relation to feelings, roots, function, doors, objects and bases [33]. Each thought manifests the feeling of pleasure (*sukha*), pain (*dukkha*), joy (*somanassa*), displeasure (*domanassa*) or equanimity (*upekkha*). Each thought is also rooted in greed, hatred, delusion, non-greed, non-hatred or non-delusion. Thoughts have one or more functions (14), namely, rebirth-linking, life continuum, adverting, seeing, hearing, smelling, tasting, touching, receiving, investigating, determining, 'javana', registration and death. There are also doors, namely, eye door, ear door, tongue door, nose door, body door and mind door, through which each thought is created. Any thought is also associated with an object, namely, visible object, sound object, smell object, taste object, tangible object and mental object. The base is the physical support for the occurrence of thoughts. Note that base is different from doors, which are the channels for formation of thoughts. *Abhidhamma* also defines many other concepts including thought processes, description of matter, and relationship between mind and matter.

4 Computational ontology of mind

This section describes the design and implementation of ontology of mind from Buddhist philosophical system. Primarily, we have used Protégé toolkit [21] for design and development of the ontology. We have named our ontology as OntoBM, an acronym for **Onto**logy of **B**uddhist theory of **M**ind.

4.1 Design of OntoBM

OntoBM has been designed with the principle that the mind can be modeled effectively through 52 mental factors as the basis for the whole modeling process. Thus, thoughts can be formed as permitted combinations of mental factors. At the top level, three major categories of mental factors, namely, ethical variable, unwholesome and wholesome have been modeled in OntoBM. The details of thoughts as feelings, roots, function, doors, objects and bases have also been annotated with each thought. OntoBM has been designed to be able to annotate the ontology at any time with further details of any node in the ontology. Furthermore, OntoBM is designed to connect with ontology query engine or applications such as multi-agent systems. Ontology Web Language (OWL) was used to write OntoBM into Protégé. Figure 1 shows a portion of the ontology of mental factors within Protégé environment.

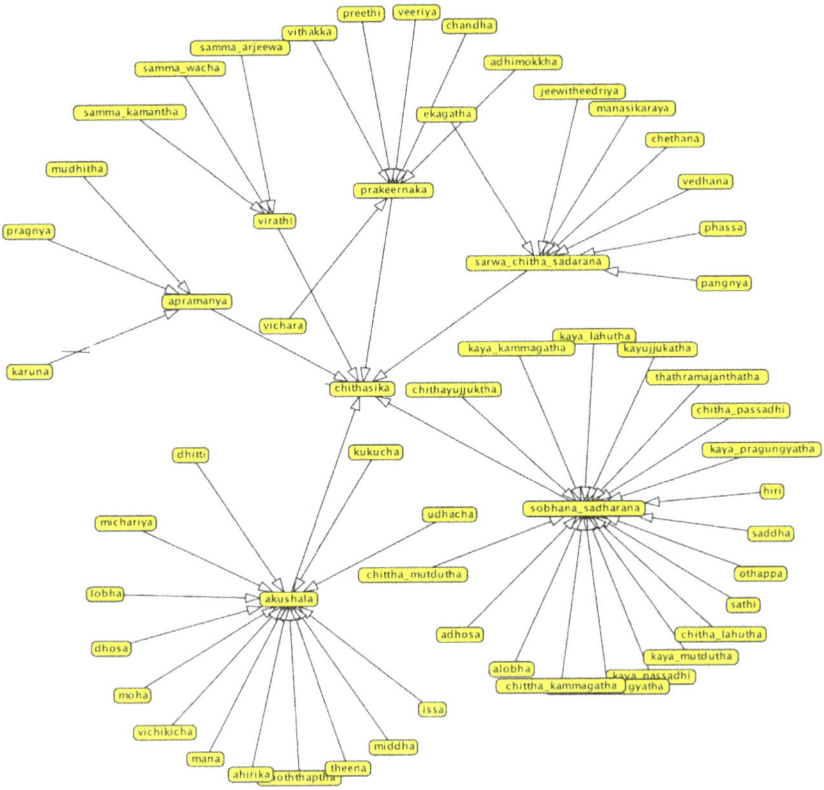

Figure 1: Ontology of mental factors.

4.1.1 Reasoning on OntoBM

Ontology being an inherently complex structure, reasoning on OntoBM is also a challenging task. To develop a reasoner on OntoBM, we used OWL API [35]. Within the context of OWL API, there are many OWL reasoners including Chainsaw [24], FaCT++ [25] and Pellet [26]. We have chosen Pellet reasoner because it is efficient in reasoning and querying. Also, it is freely available for downloading and readily compatible with Protégé environment.

4.1.2 OntoBM interface design

Graphical user interface of OntoBM has been designed to answer queries under different experimental conditions. The current interface of OntoBM primarily provides two views as Mental Factor View (MFV) and Thought View (TV). MFV allows to study or research by examining through mental factors. This view also allows the examining effect of modulation of mental factors on thoughts. In contrast, TV allows interaction with OntoMB through various thoughts as the inputs. Figure 2 depicts the MFV on OntoBM.

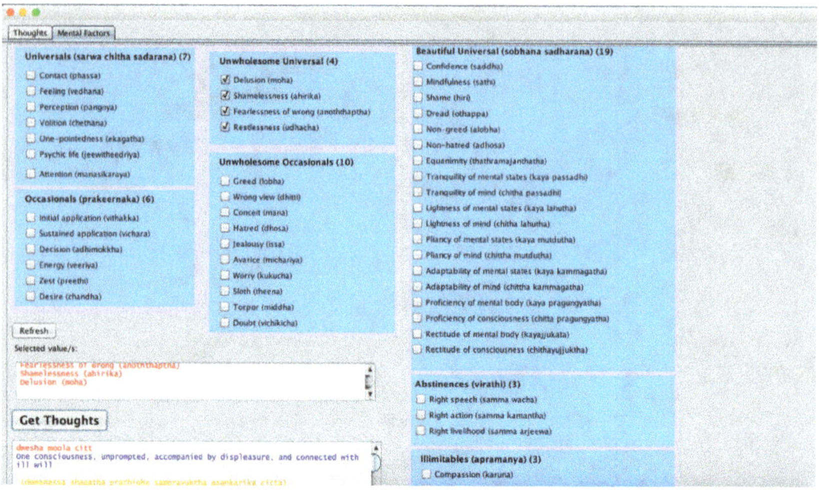

Figure 2: Mental factor view.

5 Using ONTOBM

OntoBM primarily allows learners and researchers to choose arbitrary combinations of mental factors to investigate whether such combinations could form a set of valid thoughts. When a user attempts to choose illegal (undefined) combinations of mental factors, for instance, greed with hatred, OntoBM does not allow continuing. On the other hand, when selecting a mental factor, OntoBM automatically sets the other mental factors that co-exist with the already selected mental factors. Thus, the OntoBM interface has some smart features, so that a user with zero knowledge in Buddhist theory of mind can still use the system. It should also be noted that even a person who knows *Abhidhamma* could not remember all associations of thoughts and mental factors off hand. Thus, OntoBM will be a learning tool as well.

Obviously, a knowledgeable researcher can introduce experimental combinations of mental factors and examine how the test case effects the formation of thoughts. If the selected combinations of mental factors cannot form any thoughts, the reasons will also be displayed by OntoBM. A researcher may also examine the mental factor which occurs in the largest number of unwholesome thoughts and which wholesome mental factor can counter it. In addition, a user can also select a thought and find its details such as mental factors, functions and bases. Further annotations to the OntoBM are still in progress. Depending on the researcher's creativity, many experiments can be done with OntoBM.

6 Conclusion

This paper describes a computational modeling of thought according to the *Abhidhamma* system in Buddhist psychology. We presented our research on the development of a computer model of mind from a Buddhist philosophical perspective of mind. This research has exploited *Abhidhamma* of Theravada Buddhist theory of mind to model the mind in terms of 52 mental factors, 89 thoughts and other related features of thoughts. Based on *Abhidhamma,* we have developed a computer model of ontology of mind, named as OntoBM. We have illustrated with MFV and TV, which can be used to research into behavior of mental factors and thoughts. OntoBM can be primarily used to examine combinations of mental factors that can form thoughts, associations among mental factors and details of thoughts. More importantly, researchers and scholars can also use OntoBM to experiment with the effect of modulation of mental factors on the formation of thoughts. As such, OntoBM can be used by many individuals including learners of Buddhist model of mind, researchers in brain–mind studies, meditation practitioners/teachers/researchers and psychologists. While in the development process, the results of experiments on OntoBM tested against the *Abhidhamma* text is encouraging. As further work, we are working on the development of a versatile mind simulator based on the ontology of mind, OntoBM, presented in this paper.

References

[1] Rzevski, G. & P. Skobelev, *Managing Complexity*, WIT Press, 2014.

[2] Dworschak, M, Neurotechnology: Growing a Brain in Switzerland, *Spiegel Online*, 16 Feb. 2007.

[3] B.C.Q.C., Live Science Contributor, Uploading the Mind: Could a Digital Brain Feel Pain? *LiveScience.com.* [online]. Available: http://www.livescience.com/46052-mind-uploading-ethics.html [accessed: 06-Nov-2014].

[4] Markram, H, The Blue Brain Project, *Nat. Rev. Neurosci.*, **7(2),** pp. 153–160, Feb. 2006.

[5] Brain simulator, *Softonic*. [Online]. Available: http://en.softonic.com/s/brain-simulator [accessed: 09-Nov-2014].

[6] Cao, M., Wang, J.-H., Dai, Z.-J., Cao, X.-Y., Jiang, L.-L, Fan, F.-M., Song, X.-W., Xia, M.-R., Shu, N., Dong, Q., Milham, M.P., Castellanos, F.X., Zuo, X.-N, & He, Y. Topological organization of the human brain functional connectome across the lifespan. *Dev. Cogn. Neurosci.*, 7, pp. 76–93, Jan. 2014.

[7] Campbell, I.G., EEG Recording and Analysis for Sleep Research. *Curr. Protoc. Neurosci. Editor. Board Jacqueline N Crawley Al*, Chapter 10, Unit10.2, Oct. 2009.

[8] Priori, A., Brain polarization in humans: a reappraisal of an old tool for prolonged non-invasive modulation of brain excitability. *Clin.*

Neurophysiol. Off. J. Int. Fed. Clin. Neurophysiol., **114(4)**, pp. 589–95, 2003.

[9] Steward, H., *The Ontology of Mind: Events, Processes, and States*. Clarendon Press Oxford Philosophical Monographs, 1997.

[10] Stoljar, D., & Steward, H., The Ontology of Mind: Events, States and Processes. *Philos. Rev.*, **108(3)**, p. 418, July 1999.

[11] Karunananda, A.S., Using an eastern philosophy for providing a theoretical basis for some Heuristics used in artificial neural networks. *Malays. J. Comput. Sci.*, **15(2)**, 2002.

[12] Karunananda, A., Computer Modelling of thought processes. *Int. J. Comput. Appl. Technol.*, **6(2-3)**, pp. 135–140, 1993.

[13] Narada, M.T., *A Manual of Abhidhamma*. Buddhist Missionary Society, 1979.

[14] Prigogine, I., *The End of Certainty*, 1 edition. New York: Free Press, 1997.

[15] Prigogine, I., *Is Future Given?* World Scientific, 2003.

[16] Herculano-Houzel, S., The remarkable, yet not extraordinary, human brain as a scaled-up primate brain and its associated cost. *Proc. Natl. Acad. Sci.*, **109(1)**, pp. 10661–10668, June 2012.

[17] Dave Evans, The Internet of Things How the Next Evolution of the Internet Is Changing Everything. 2011.

[18] Gruber, T.A., A Translation Approach to portable Ontology Specification. *Knowl. Acquis.*, **5(2)**, pp. 199–220, 1993.

[19] Luciano, J.S., Andersson, B., Batchelor, C., Bodenreider, O., Clark, T., Denney, C.K., Domarew, C., Gambet, T., Harland, L., Jentzsch, A., Kashyap, V., Kos, P., Kozlovsky, J., Lebo, T., Marshall, S.M., McCusker, J.P., McGuinness, D.L., Ogbuji, C., Pichler, E., Powers, R.L., Prud'hommeaux, E., Samwald, M., Schriml, L., Tonellato, P.J., Whetzel, P.L., Zhao, J., Stephens, S. & Dumontier, M., The Translational Medicine Ontology and Knowledge Base: driving personalized medicine by bridging the gap between bench and bedside, *J. Biomed. Semant*, **2(2)**, p. S1, May 2011.

[20] Rodriguez-Rocha, B.D., Castillo-Barrera, F.E. & Lopez-Padilla, H., Knowledge Capitalization in the Automotive Industry Using an Ontology Based on the ISO/TS 16949 Standard, in *Electronics, Robotics and Automotive Mechanics Conference*, Los Alamitos, CA, USA, 2009, pp. 100–106.

[21] Protégé. http://protege.stanford.edu, 2012.

[22] Noy, N.F. & Musen, M., PROMPT: Algorithm and tool for automated ontology merging and alignment, *Proc. 17th Natl. Conf. Artif. Intell. AAAI-2000 Austin TX*, 2000.

[23] Acampora, G., Avella, P., Loia, V., Salerno, S. & Vitiello, A., Improving ontology alignment through memetic algorithms, in *2011 IEEE International Conference on Fuzzy Systems (FUZZ)*, pp. 1783–1790, 2011.

[24] Tsarkov, D. & Palmisano, I., Chainsaw: A Metareasoner for Large Ontologies. in *CEUR Workshop Proceedings*, 2012.

[25] Abburu, S., A Survey on Ontology Reasoners and Comparison. *Int. J. Comput. Appl.*, **57(17)**, 2012.
[26] Sirin, E., Parsia, B., Grau, B.C., Kalyanpur, A. & Katz, Y., Pellet: A practical OWL-DL reasoned. *Web Semant. Sci. Serv. Agents World Wide Web*, **5(2)**, pp. 51–53, June 2007.
[27] Wooldridge, M., *An Introduction to MultiAgent Systems*. John Wiley & Sons, 2009.
[28] Rzevski, G., Modelling Large Complex Systems Using Multi-agent Technology, pp. 434–437, 2012.
[29] Hongqiao, Y., Xihua, L., Fei, W. & Weizi, L. Multi-agent based modeling and simulation of complex system in hospital, in *16th International Conference on Industrial Engineering and Engineering Management, 2009. IE EM '09*, pp. 1759–1763, 2009.
[30] Belgacem, S. & Fates, N., Robustness of Multi-agent Models: The Example of Collaboration between Turmites with Synchronous and Asynchronous Updating, *Complex Syst.*, **21**, 2012.
[31] Singh, A., Juneja, D. & Sharma, A.K., Agent Development Toolkits, *Int. J. Adv. Technol.*, **2(1)**, pp. 158–164, 2011.
[32] Nyanaponika, T., *Ahidhamma Studies – Research in Buddhist Psychology*. Wisdom Publications, 2013.
[33] Bodhi, B., *Abhidhammatha Sangaha: a comprehensive mannula of Abhidhamma*. Onalaska: BPS Pariyatti Edition, 2000.
[34] Kalupahana, D.J., *The principles of Buddhist Psychology*. State University of New York Press, 1987.
[35] Horridge, M. & Bechhofer, S., The owl api: A java api for owl ontologies, *Semantic Web*, **2(1)**, pp. 11–21, 2011.

Complexity as the defining feature of the 21st century

G. Rzevski[1,2]
[1]*The Open University, UK*
[2]*Milton Keynes & Multi-Agent Technology Ltd., UK*

Abstract

The 21st century is often labelled as the Century of Complexity because complexity of the socio-technological environment in which we live and work has reached the level that no longer can be ignored – complexity affects now all aspects of our lives. The paper defines complexity, identifies its sources and explains the basics of the new science and the art of managing complexity. The importance of complexity is discussed in the context of social, business and engineering systems.
Keywords: adaptation, complexity, managing complexity, multi-agent technology, real-time decision-making, self-organisation.

1 Introduction

Complexity is an inherent property of many systems that constitute the environment in which we live and work, namely, ecological, biological, thermodynamic and social systems (including political, administrative, economic, business and socio-technical systems). Until recently, the levels of complexity of social systems were low and, consequently, complexity was largely ignored. However, with the rapid development of digital technology, the situation has changed, particularly when the Internet transformed the world into a 'global village' and linked regional and national markets into a single 'global market'.

Many researchers have contributed to the understanding of complexity, notably Prigogine [1, 2], Kaufman [3], Holland [4] and many others. As always at the onset of a paradigm shift, there are several competing 'schools' of complexity science.

WIT Transactions on State of the Art in Science and Engineering, Vol 90, © 2016 WIT Press
www.witpress.com, ISSN 1755-8336 (on-line)
doi:10.2495/978-1-78466-155-7/003

This paper is based on the pioneering work on developing experimental science and the art of managing complexity [5].

2 What is complexity?

Complexity is a property of open systems that consist of a large number of diverse, interacting components, often called agents. Complex systems can be distinguished from other systems by the seven features: connectivity, autonomy of agents, emergent behaviour, nonequilibrium, nonlinearity, self-organisation and co-evolution.

Connectivity – Agents are interconnected. Complexity of the system increases with the number of links that connect agents to each other. Complexity also depends on the strengths of links. The weaker the links between agents, the easier is to break them and form new ones, which increases system complexity.

Autonomy of agents – Agents have certain freedom of behaviour (autonomy), which is always limited by norms, rules, regulations and/or laws. The increase in autonomy of agents increases the complexity, and if all constraints on agent behaviour are removed, the system switches from complex to random behaviour. Inversely, if autonomy of agents is reduced (by tightening of laws and regulations), the system complexity will decrease, and in the extreme, the system will become deterministic. Complex systems have no central control.

Emergent behaviour – Behaviour of complex systems emerges from the interactions of agents and is not predictable, and yet it is not random. Uncertainty about the outcome of agent interactions is always between 0 and 1.

Nonequilibrium – Complex systems are subjected to perpetual change experienced either as a succession of discrete disruptive events or as a slow, imperceptible drift into failure. Frequency of disruptive events varies with complexity. In systems of high complexity, disruptive events occur so frequently that the system has no time to return to stable equilibrium before the next disruption occurs. When complexity levels are very high, the system is said to be at the edge of chaos because the uncertainty of behaviour is close to 1.

Nonlinearity – Relations between agents are nonlinear. Nonlinearity may amplify a small, insignificant disruptive event and cause a catastrophic outcome (an extreme event), the property called butterfly effect. Butterfly effect increases with complexity. In complex systems, the outcomes are, as a rule, consequences of numerous interacting causes, and therefore, the cause–effect analysis is inappropriate.

Self-organisation – Complex systems have a propensity to react to disruptive events by autonomously self-organising with the aim of eliminating or, at least, reducing the consequences of disruption, the property called adaptation. Self-organisation may be also caused autonomously by a propensity to improve own performance, the property called creativity or innovation. To initiate and perform adaptive and creative activities, the system must be intelligent. Intelligence, adaptation and creativity are the properties exclusive to complex systems and their levels increase with complexity.

Co-evolution – With time, complex systems co-evolve with their environments. Co-evolution is irreversible.

3 Complex versus deterministic and random systems

Let us use uncertainty of behaviour as the demarcation parameter to distinguish complex systems from deterministic and random systems, as shown in Table 1.

The term deterministic implies that uncertainty is equal to 0, while the term random means that uncertainty is equal to one. Complex systems have uncertainty value between 0 and 1.

Table 1 highlights the link between complexity and uncertainty of system behaviour: uncertainty is a consequence of complexity and it increases with complexity. Low complexity systems have uncertainty close to 0 and their behaviour differs little from the behaviour of deterministic systems. Highly complex systems with uncertainty close to 1 are at the edge of chaos and their behaviour is characterised by features such as self-organisation, generation of unpredictable extreme events and co-evolution.

Table 1: Complex versus deterministic and random.

Random	Complex	Deterministic
Uncertainty of outcome = 1	1 > Uncertainty of outcome > 0	Uncertainty of outcome = 0
Components have full autonomy	Components (called agents) have partial autonomy	Components have no autonomy
Disorganised	Self-organising Evolving	Organised
Unpredictable behaviour	Emergent behaviour	Predictable behaviour

The distinction between complex and deterministic systems is very important and has philosophical repercussions. For centuries, eminent philosophers and scientists have believed that the world is deterministic – that it behaves in accordance with natural laws in a predictable manner and that any uncertainty of outcomes is a result of our lack of knowledge on how the world works. In other words, for supporters of determinism, the world is complex only for those who do not understand it.

A more plausible alternative view has been put forward recently by Prigogine [1, 2]. The world is inherently complex and it evolves with time. Future is not given; it emerges from the interaction of billions of activities performed by constituent agents, including people, animals, plants as well as natural forces such as climate, erosion, volcanic eruptions and solar spots. Only certain limited parts of the world can be represented by deterministic models such as Newton's laws.

4 Co-evolution of society and technology

Historically, complexity of social systems increased in steps, driven by the advances in technology, as depicted in Table 2, and at each step, the impact on the way we live and work was dramatic.

During the transition between agricultural and industrial societies, rapid migration of the population from the countryside to the cities, to take advantage of new employment opportunities, increased the social connectivity in the increasingly dense cities and, as a result, raised the level of social complexity. The massive movement of population caused well-documented disturbances as a rigid, traditional social order based on land ownership was replaced by a chaotic transition, which then settled into a new social order based on ownership of capital.

The current transition from industrial to information society, which began after the end of the World War 2 with the invention of computers, is particularly notorious by the very steep increase in social complexity caused by the rapid spread of digital technology, which offers unparalleled social connectivity (social density), but this time without any need for the population to move. Now we can form communities of interests across the globe. Distances do not matter anymore.

Thanks to digital technology, participants in the information society interact faster, more frequently and with greater number of correspondents than ever before. In the year 2013, approximately 3 billion people have used the Internet, which is more than 40% of the total number of people on the planet and, according to the Time news feed, up to 6 billion people had access to a mobile phone. This is an astonishing increase in connectivity causing an accelerated growth of complexity.

Table 2: Co-evolution of technology and society.

Social stages	Key resources	Distribution	Scope	Success factors
Information society supported by digital technology	Knowledge	Digital networks	Global	Adaptability
Industrial society supported by mass production technology	Capital	Railways motorways	National	Economy of scale
Agricultural society supported by manual labour	Land	Local roads	Local	Hard work

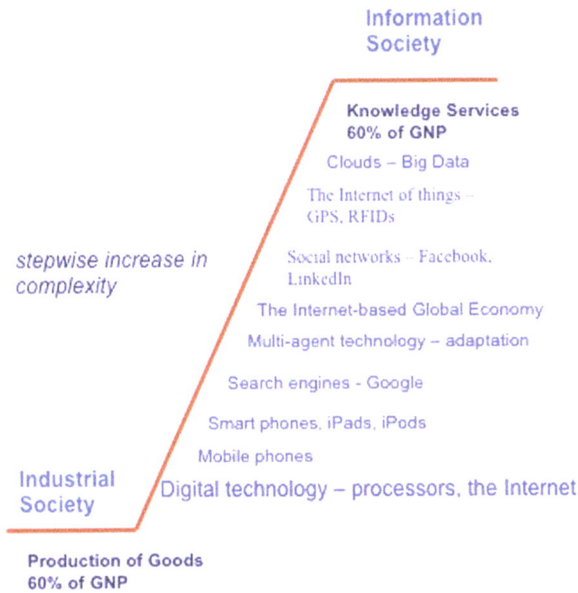

Figure 1: Steep increase in complexity of social systems caused by rapid penetration of digital technology.

As connectivity has increased, we have experienced a very important shift from nation-centred industrial markets to the global economy dominated by knowledge-based services. In the industrial economy, money could buy any knowledge needed for business. In the knowledge economy, the knowledge of how to solve complex problems can attract investments that are required to start and sustain an economic activity. Pioneers of the knowledge economy, founders of knowledge-based companies such as Apple, Google, Amazon, Microsoft and Facebook, are the new economic elite. The shift of mass manufacturing from the developed to the developing countries is a part and parcel of globalisation. However, the replacement of mass manufacturing by knowledge-based services as the main wealth creation activity occurred only in the countries where there exists expertise in advanced IT and a large number of high-class knowledge workers: researchers, designers and decision-makers in financial services, IT, engineering, consulting, construction, architecture, entertainment, media, etc.

Big monolithic corporations are the product of the industrial economy, which was characterised by stable markets generating steady demands for identical, mass produced goods. Big corporations were designed to be rigid and permanent and they thrived in the era when the Economy of Scale was the key success factor. The new complex global market is the enemy of anything big and rigid. The new critical success factor is adaptability, and therefore, we can safely assume that large corporations will not have an easy future, with the exception of those that manufacture uniform products exemplified by nappies or nuts and bolts.

However, big corporations have a remarkable ability to survive and many will re-invent themselves and continue in a more appropriate format. The organisational structure that is the most suitable for delivering to perpetually changing markets is a network of self-contained production units, each having a unique expertise (knowledge resource), often referred to as Virtual or Digital Enterprise.

Figure 2: The global network.

The concentration of data on financial transactions, on communication with friends and business associates, and on individual mobility in huge 'clouds', by organisations such as Google, raises important questions on individual privacy. It is only natural to expect that those who have knowledge about us will try to use this knowledge to manipulate our behaviour. Knowledge is power. Who will exercise this power acquired by accumulation of digital data about every aspect of our life? Will a private company (possibly in collusion with a government intelligence service) manage to acquire sufficient quantity of data to establish monopoly of knowledge? Or, can we expect that the process of natural selection will ensure the distribution of knowledge? It is safe to be an optimist. Evolution favours complexity, which implies diversity and distributed decision-making rather than centralisation, although the process is slow and by no means smooth.

Figure 2 shows the emerging global network; as all texts, images and videos/films are digitised (the Internet of Documents), more and more people are connecting through mobile devices and the Internet (the Internet of People) and objects of practical importance to humans are furnished with electronic tags, enabling them to communicate with each other bypassing their users (the Internet of Things).

5 Prospering in a complex environment

We have been brought up in Newtonian deterministic tradition and feel comfortable in well-defined stable situations. We crave for simplicity and predictability.

As it happens, complexity of the environment in which we live and work is relentlessly increasing and it now intrudes into every aspect of our existence. The increase in complexity is disruptive – by making our well-established systems and processes ineffective, it forces paradigm shifts *opening up opportunities for creating a new order in society, politics, law, policy, education, research, business, design, engineering and elsewhere.*

To take advantage of new opportunities, it is helpful for individuals and organisations to develop the so-called 'complexity mindset', which, in a nutshell, consists of beliefs, principles and methods that define the relation between an individual or an organisation and the ever-changing world with which they have to co-evolve.

6 Managing complexity

A new scientific discipline entitled 'managing complexity' contains a growing collection of concepts, principles and methods for successfully living and working with complexity [5].

There are two aspects of managing complexity:

- Coping with external complexity (complexity of the environment) and
- Creating and tuning internal complexity

Some of the key concepts and principles are briefly outlined below.

6.1 Coping with external complexity

By definition, we do not have control over our environment, and therefore, we cannot control its complexity. The best strategy for coping with external complexity is to develop the capacity for adaptation, which implies designing complexity into our processes and structures because only complex systems can self-organise and thus adapt.

To be adaptive means to be able to achieve desired goals under conditions of frequent occurrence of unpredictable disruptive events. Adaptability is achieved by rescheduling affected resources to eliminate or, at least, to reduce consequences of a disruptive event before the next one occurs.

Key requirements for adaptability are:

- Distributed rather than centralised decision-making
- A sufficient redundancy of resources to enable unpredictable rescheduling
- Availability of technology capable of
- Early detection of disruptive events
- Real-time rescheduling of affected resources
- Continuous improvement of performance to avoid a drift into failure

6.2 Tuning internal complexity

The level of complexity of systems/organisations, which we design or control, can be adjusted by changing

- Agent autonomy and
- Agent connectivity

This is largely a trial-and-error process, informed by experience in designing and managing large-scale complex adaptive systems, as described in some detail in [5].

7 Complexity and society

For the purposes of this paper, I shall assume that any system in which large numbers of agents are people is a social system. Then we can postulate that any social system in which members (a) have a high level of autonomy, (b) are able to communicate with many other members and (c) can change established communication links easily is a complex social system. Such a system is adaptive and, therefore, resilient to disruptions and attacks; it co-evolves with its environment, and therefore, it never becomes obsolete [6].

And yet, we tend to design organisations that are pseudo-deterministic and controlled by elaborate hierarchical management structures, rather than self-organising. To achieve this, we constrain human natural intelligence and creativity.

8 Complexity and business

A business is of course a social system. It is necessary though to single out peculiarities of this class of social systems because businesses create wealth [5].

A great majority of businesses today participate in the Internet-based global market. Suppliers, customers, traders, investors, bankers, consultants and middlemen are negotiating, agreeing, modifying or cancelling transactions in unprecedented numbers and with unprecedented speeds. As a consequence, the global market has become volatile and the frequency of disruptive events is such that the market, once disturbed, has no time to return to stable supply/demand equilibrium. Complexity of the global market cannot be reduced or influenced in any way by any business, even by the biggest one. Also, it has to be accepted that the complexity will not go away. In fact, there exists ample evidence that it will increase as more and more people and things connect to the Internet.

Businesses operating under such conditions have considerable difficulty in planning and managing their internal business processes. Traditional decision-making and resource optimisation approaches do not deliver expected results.

Clearly, the solution is to develop capacity for adaptability.

Manufacturing of cars is an instructive example. One of the leading European car manufacturers, for whom I worked as a consultant, still spends a large sum of money on out-of-date production optimisation software that requires 8 h or so to produce a perfect (deterministic) production schedule, which becomes obsolete a couple of hours after it is implemented because changes to previously agreed orders arrive from dealers approximately one per 2 h. As a result, the company manufactures cars that are not wanted anymore and have to be sold with discount.

A real-time, complex adaptive scheduler, such as one described in [6], would cost less and would be able to re-schedule affected parts of the production plan whenever a request for a change arrives.

9 Complexity and engineering

Engineering systems have been always designed to exhibit deterministic behaviour within specified operational range. To achieve this aim, dynamic engineering systems that have a variety of possible behaviours, such as machines, vehicles, aircraft, rockets and robots, had to be provided with a controller, human or automatic, to behave predictably.

In contrast, the behaviour of complex systems is emergent (unpredictable) and these systems have no controllers – they are adaptable – and they self-organise when disturbed to eliminate or, at least, to reduce the consequences of the disturbance. This is precisely why complex engineering systems are valuable when they operate in complex environments where adaptation has an advantage.

Consider an example. The US sent a robot to Mars, which stopped functioning after a few weeks because space dust covered its solar cells. Then UK sent a robot to Mars, which fell into a crevice before it could even start to do some useful work. That was a massive waste of resources and time entirely due to outdated thinking of robot designers. A family of five much smaller robots, as shown in Fig. 3 below, would be able to share the workload, clean and maintain each other, and either drag out the unfortunate family member that fell into a crack or abandon it and share its workload among those that survived. That would be a perfect complex adaptive engineering system.

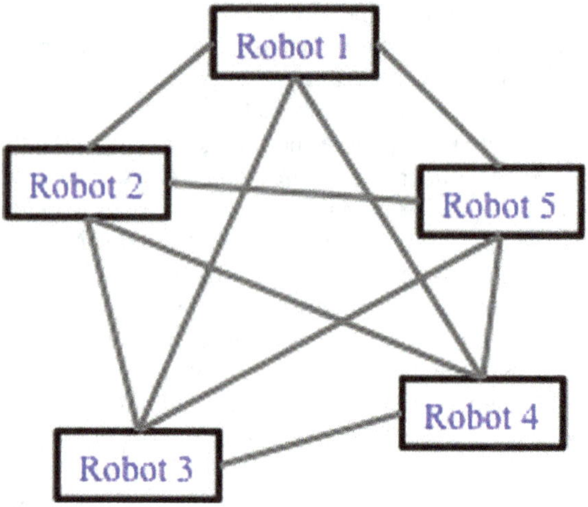

Figure 3: A complex engineering system depicting a family of five robots.

10 Conclusion

There exists ample evidence that, driven by the rapid development of digital technology, complexity of our socio-economic environment is perpetually increasing and it has by now reached the level that no longer can be ignored.

It is, therefore, of paramount importance for all those who live in the new global village and work in the emerging global market to develop the appropriate complexity mindset, which would enable them to take advantage of fresh opportunities in research and practical applications currently on offer.

References

[1] Prigogine, I., *The End of Certainty: Time, Chaos and the new Laws of Nature*, Free Press, 1997. ISBN 0-684-83705-6.

[2] Prigogine, I., *Is Future Given?*, World Scientific Publishing Co., 2003. ISBN 981-238-508-8.

[3] Kaufman, S., *At Home In the Universe: The Search for the Laws of Self-Organization and Complexity*, Oxford Press. 1995. ISBN 0-19-511130-3.

[4] Holland, J., *Emergence: from Chaos to Order*, Oxford University Press, ISBN 0-19-850409-8. 1998.

[5] Rzevski, G., Skobelev, P., *Managing Complexity*, WIT Press, 2014. ISBN 978-1-84564-936-4.

[6] Rzevski, G. *Self-Organisation in Social Systems*, Ontology of Designing. **4(14)**, pp. 8–17, 2014. ISSN 2223-9537.

[7] Rzevski, G., *Using Complexity Science Framework and Multi-Agent Technology in Design*, In *Embracing Complexity in Design*, eds. K. Alexiou, J. Johnson, T. Zamenopoulos, Routledge, 2010, ISBN 978-0-415-49700-8, pp. 61–72.

Factors that facilitate organisational change in complex systems

P. M. Dickens
Tyndale University College, Canada

Abstract

This paper explores the capacity of complex systems to find their own form of order and coherence, often referred to in terms of self-organisation emergent change, then asking the question, 'What can organisational leaders do to create the systems and structures that would facilitate emergent change?' Emergent change comes from within and through the active members of a system and is not imposed according to some external prompting or design. This results in the sort of change capacity that enables an organisation to be agile and resilient in highly volatile times. I have identified seven key organisation-specific factors that facilitate emergent change. These include: executive engagement, purposeful orientation, a culture of experimentation, a safe-fail culture, collaborative decision-making, collaborative quality measures, and intentional learning. These factors were initially identified through an extensive literature review, interviews with the CEOs of 15 Canadian healthcare organisations, and a think tank of subject matter experts. This resulted in the use of exploratory factor analysis to validate a survey that can be used to assess the presence or absence of these factors in a specific organisation, thus providing leaders with a framework for change.
Keywords: change, collaboration, creativity, emergence, leadership, learning, self-organisation.

1 The challenge

As organisational life becomes more and more complex, today's dominant management paradigms no longer seem to suffice. The problem is that while many leaders readily acknowledge the challenges and limitations of traditional approaches, they have a limited range of options for dealing with a highly

WIT Transactions on State of the Art in Science and Engineering, Vol 90, © 2016 WIT Press
www.witpress.com, ISSN 1755-8336 (on-line)
doi:10.2495/978-1-78466-155-7/004

complex world. This paper identifies new ways of thinking about and responding to complexity: not by trying to simplify it, but by embracing the inherent capacity of complex systems to find their own form of order and coherence.

The level of complexity and unpredictable change faced by organisations today is unprecedented. Globalisation, rapidly changing technologies, unpredictable geopolitical environments, and increasingly informed consumers bring tremendous challenges for organisations trying to stay afloat in such 'permanent white water' [1]. Eoyang and Holladay [2] suggest that we can best understand this complexity along three continua: bounded and unbounded systems, few factors and many factors, and one-way and two-way causality. When the problems we face are relatively simple, the system within which we operate is described as bounded or contained. We focus on the implications of a decision on our organisation or department. A problem becomes more complicated when there are multiple systems impacted by any one decision, and a problem is truly complex [3] when we realise there are almost innumerable systems impacted by a change (Zimmerman *et al.* [3] provide a useful series of metaphors to understand the difference between simple, complicated, and complex. Baking a cake is simple: the recipe is known, tested, and replicable. Launching a rocket is complicated: multiple formulae, careful attention to sequence, but ultimately replicable. Raising children is complex: formulae are of limited value, no two are alike, and the results are ultimately unpredictable). Similarly, the exponential increase of factors impacting a single decision moves the marker along the second continuum and the presence of multiple causalities makes more and more challenges extremely complex. Little wonder that Eoyang suggests that, 'well-informed trial and error is the only viable strategy' [2, p. 20] in today's highly complex world. Hence, I see an urgent need to develop a better understanding of complexity and complex, adaptive systems (CAS) from a leadership perspective.

2 Broad definitions of CAS

The origins of complexity theory lie in the study of deterministic systems such as biological units, numbers, and subatomic particles – in other words, systems that lack free will and choice. However, the reality for leaders is that theirs are human-based systems and, as such, are subject to the inherent complexities of both the human body and, perhaps more challenging, the human mind and will. There are multiple definitions of CAS. According to Plowman *et al.* [4], some of the characteristics of CAS include: (1) they are made up of many agents who act and interact with each other in unpredictable ways; (2) they are sensitive to initial conditions; (3) they adjust their behaviour in the aggregate in unpredictable ways; (4) they oscillate between stability and instability; and (5) they produce emergent actions when approaching disequilibrium [4].

These descriptors of various attributes of CAS are central to the factors that this paper identifies as key to facilitating emergent change. What this paper tries to do is to take the discussion of attributes further by trying to assess the presence or absence of such factors, in order to provide leaders with suggested strategies to enhance their organisational capacity for change.

In a complex, adaptive social system, no individual agent, or designated group of agents, determines the local interaction principles of others and there is no centralised direction managing the patterns of behaviour of the system as a whole or of the evolution of those patterns. The evolution is, in fact, referred to as self-organisation. It is important to recognise that conceptually, emergent change is neither positive nor negative: it simply is. A riot is as much an example of self-organisation and emergent change as was the gathering that heard Martin Luther King's famous 'I have a dream' speech. It is my belief that we can distinguish positive emergent change in a social system because positive emergent change contributes to the mutual purposes of both leaders and followers who share common values and intend real change [5].

Many researchers have suggested that self-organisation is the key concept to be drawn from complexity theory.

> Self-organisation is a characteristic of all social systems that operates whether we recognise it or not. By recognising it, however, we can begin to influence it to facilitate better outcomes. Self-organisation is the process by which people mutually adjust their behaviours in ways needed to cope with changing internal and external environmental demands [3, 6–9].

Self-organisation is a process whereby the coherence of a system spontaneously increases, without this increase being controlled by formal directive or an external system. This spontaneous increase in order is a construct that would stand in marked contrast to the concept of entropy described in classic Newtonian physics. Some would argue that there is enormous benefit to organisations that can develop the capacity for self-organisation. As Lindberg *et al.* suggest, 'The more self-organised the change (in an organisation), the higher the whole system's performance will be' [10]. That being said, it does not provide a guarantee. In the same way, certain parenting approaches increase the likelihood of a well-adapted child, there are a great many other intervening variables that may prevent this outcome from happening.

3 Research methods

The research was based on a review of the literature, interviews with CEOs in 15 large, complex health care organisations, feedback from a panel of subject matter experts, and my own organisation development practice experience. This allowed me to identify nine initial constructs of interest, which were then integrated into the initial Organisational Resilience Assessment (ORA). One hundred and sixty-two senior leaders, managers, and staff at a hospital in Toronto, Canada completed the survey. I then completed a principal component analysis of the constructs to validate the survey. From this analysis, seven valid factors emerged. I assessed the link between the factors and organisational performance by conducting two focus groups at the study site. The focus groups confirmed the correlation between the factors and key organisational performance indicators. The factors include: purposeful orientation, a high level of executive engagement, a safe-fail culture, collaborative decision-making processes, collaborative quality processes, intentional learning process, and culture of experimentation.

3.1 Initial interviews

Prior to the development of the ORA survey, I interviewed senior leaders in the Ontario healthcare sector. This included 13 hospital CEOs, a Deputy Assistant Minister in the Ministry of Health and Long-Term Care, and a senior researcher. There was broad consensus that the system is extremely complex and that the complexity is increased by the lack of clarity regarding the role of the Local Health Integration Networks (LHINs), Ontario's initial foray into some form of regional delivery model. Broadly speaking, all the interviewees pointed to the importance of several of the elements of the seven validated factors that emerged from the exploratory factor analysis, particularly the importance of leadership engagement and organisation-wide leadership development in order to provide the skills to engage in the sorts of collaborative behaviours indicated by the survey results. The feedback from the interviews aligned with the information drawn from the literature review thus forming the basics of the initial constructs of interest for the survey.

3.2 Development and evaluation of the ORA survey

The survey was developed following a careful analysis of the literature and the interviews, looking for common patterns and themes. From this analysis, seven initial constructs of interest were developed. A panel of individuals with some experience with emergent change was then asked to identify specific factors that they felt had facilitated the change. From this feedback, two new constructs emerged, as well as the language that began to frame items within each construct. A draft of the survey was then circulated to the panel for further refinement. Once the draft had gone through two more iterations, an Ethics Review was completed at the study site. The site was selected because it consistently achieved strong financial, patient care, and staff satisfaction results.

The survey was conducted using Survey Monkey® and allowed for both quantitative and qualitative responses. One hundred and seventy-four leaders and staff at the hospital responded to the survey, but 10 did not complete it, leaving N = 164. The survey group was intentionally limited to past participants in a formal leadership development programme at the hospital, in the expectation that it would increase the response rate. It did (response rate was 40%), but it also appears to have skewed the results, which were consistently above the mean. Since this was the first use of the survey, there was not a pool of data against which to assess the results; but since the intent was validation of the survey rather than the formal assessment of the hospital, the process served its purpose. Quantitative data, in the form of individual comments on each of the factors, was significantly more balanced, but the overall tone of responses indicated affirmation of all seven factors.

The results were then loaded into SPSS®, a software package that facilitates an exploratory factor analysis. After careful analysis, five factors with four to six items in each factor indicated Cronbach's alpha over 0.700, which is indicative of their internal validity. Two more emerged with alpha between 0.550 and 0.700, which indicated some level of validity, but also indicated the need for further refinement. A second version of the survey was then prepared.

4 The seven factors that facilitate emergent change

This study undertook the task of understanding and validating organisational or strategic factors that might facilitate positive emergent change. What emerged from the results of the study were as follows:

4.1 Executive engagement

The degree to which senior management demonstrates support for and commitment to non-hierarchical approaches. This had a very strong internal validity and aligned with the input of the interviews. It was later validated by the focus groups. The importance of commitment and support from senior leadership is almost counterintuitive when thinking in terms of CAS, where the emphasis tends to focus on a more distributed model of leadership [11]; however, a previous research at the study site hospital indicated the significance of senior-level support [13]. As one interviewee in that study said, 'I know it sounds a bit silly, but it was almost as if we needed, or wanted, mom and dad's permission to experiment'. This factor includes the level of visibility and approachability of the CEO and members of the senior executive team, and the degree to which they enable staff to fully engage in, and take ownership for, creative problem-solving.

The need for an intentional and widespread leadership development strategy was implicit in the interviews when the topic of executive engagement was discussed. The implication was that, without a commitment to distributed leadership, senior leaders were inevitably drawn into a more managerial role, which limited their ability to engage relations across the organisation. Equipping leaders at all levels of the organisation was viewed as central to developing the innovative and decision-making capacity of the organisation and supported a commitment to ongoing, non-clinical/technical learning.

4.2 Safe-fail culture

The degree to which the organisation is perceived to be comfortable with making mistakes, learning, and moving on without fear of recrimination. The idea that organisations need to create safe spaces to foster innovation has been present since organisations began to talk about innovative cultures [14]. However, historically, these safe spaces have been segregated and given terms like skunk works. Today, it would appear that the pace of change facing every function of an organisation does not often allow for the luxury of such separateness. Instead, organisations need to create and support safe holding spaces for experimentation [15]. While the survey validated the importance of this, written feedback and the focus groups indicated that, while vital, it was not universal at the study site. It appears to be contingent on the level of trust between staff and management.

4.3 Collaborative decision processes

The degree to which people throughout the organisation have timely input into the decisions that affect the work that they do. Holman *et al.* suggested that 'what keeps (a) system whole over time is a commitment to collaborative meaning making' [16]. The opportunities for such collaboration can be left to chance, but the strategic organisation takes the lead in creating collaborative work structures. Tekell *et al.* (in Holman *et al.* [16]) described collaborative work systems arising from 'a holistic design process that creates the framework for successfully changing the organisation to support collaboration and improve business results' [16, p. 437]. These collaborative structures create the space within which people at all levels of the organisations can discuss the implications of the three continua discussed earlier and engage in their own adaptive action. The movement to interprofessional education and collaborative care models in health care is evidence of this shift. The study site has committed significant time to developing collaborative care models, so there was intentionality about this factor that was indicated by the high variance from the mean exhibited in the survey results.

4.4 Collaborative quality

The degree to which decisions about quality measures and strategies are defined by the people doing the work, supported with the data they need to make decisions. The primary rationale for the inclusion of this factor comes from Wergin's study [17] in which he reported on the findings of an extensive Pew Charitable Trust study on quality. Wergin intended to identify the necessary conditions for quality in a post-secondary education environment. What he found was that it was key that quality occurred at the departmental level and with academic chairs, rather than at the most senior levels. The same is true in health care organisations: people closest to the patient have the best understanding of the necessary quality metrics.

While Wergin found that there were six necessary conditions, they were not sufficient. One other factor differentiated the really high-quality departments: evaluation policies were flexible and decentralised. Each department defined what quality meant in their context and then were held accountable for meeting that standard. 'The only institution-wide requirement was that departments include in their study an analysis of how they contribute to the mission of the institution' [17, p. 35]. Wergin found that six necessary conditions have to exist before there is sufficient trust to embrace decentralised evaluation. Based on Wergin's conclusions, a localised quality framework was deemed a distinct factor in this study. During the focus groups, several people pointed out that this approach to quality was central to the Lean Quality Improvement approaches that are a systemic part of the hospital; so, the importance of this factor was not a surprise to the participants.

4.5 Intentional learning processes

The degree to which there is both formal and informal support for both technical and relational skills and the willingness of the organisation to learn as it goes. Learning and constant adaptation are central elements of a CAS's capacity to adapt and change [18].

Even the simplest adaptive system has some purpose, namely, to perform some task. It follows that, unlike agents in deterministic systems, agents in all adaptive systems adjust their behaviour in light of its consequence for their purpose. In other words, adaptive systems learn, at the very least, in a simple single-loop manner, whereas a deterministic system does not. A great deal of attention is rightly placed on the capacity of the individual agents to learn and adapt [19] and for those same agents to engage in opportunities for shared learning [20, 21]; however, this study is focused on the *organisational* factors that contribute to emergence, so the focus here is on system structures. From that perspective, the question becomes whether or not the agents of the system are intentionally aligned to teams and groups that will facilitate learning. Several of the individual comments in the survey spoke of the value that respondents attached to the degree of commitment the organisation has made to their personal and professional development in terms of access to formal courses and workshops. However, there was also evident tension between availability and time: frontline staff, in particular, indicated that it was difficult for them to get away from their tasks to participate.

4.6 Culture of experimentation

The degree to which people are willing to try new ideas and approaches, to listen to people who think differently than they do, and to welcome innovation. Central to this type of experimentation is having real-time access to meaningful data in order to constantly adapt and change. Like learning, feedback is a central concept in CAS theory. Complex systems are open systems with feedback loops that can enhance and stimulate positive emergence when there is a steady flow of real-time information, so that the agents within the system can adjust their innovative efforts in order to increase effectiveness. A system must not only receive, process, and retain information, but also respond to and produce some form of output or new data to which other elements of the system can then respond. The difference between the sorts of feedback mechanisms needed in complex adaptive organisations is that the organisation needs to be able to 'hold multiple and sometimes conflicting representations of environmental variety, retaining in their behavioural repertoire a range of responses, each of which operates at a lower level of specificity' [22]. Change in a CAS is encouraged by increasing information flow to all parts of the organisation and then pushing the authority to do something with that information as far out into the organisation as possible [6]. The importance of access to real-time data in order to adapt and improve was strongly evident in both the literature and feedback from the focus group.

4.7 Purposeful orientation

The degree to which people feel they have a common purpose and focus and share a passion for that purpose. Wheatley suggested that, 'We need to trust that something as simple as a clear core of values and vision, kept in motion by continuous dialogue, can lead to order' [23]. This begins to suggest a different way of thinking about an organisation's strategic framework or high-level organisational commitments. This framework creates the 'bounded instability' that allows for innovation and self-organisation [19]. Wheatley went on to suggest that, 'In human organisations, a clear sense of identity – of values, traditions, aspirations, competencies, and culture that guide the operation – is the real source of independence from the environment' [23, p. 94]. The mission of the study site is central to all decisions made in the study site, and several survey respondents commented on its visual prominence and intentionality in decision-making.

5 Impact on performance: a molecular metaphor

Both the focus groups affirmed the importance of all seven factors to organisational performance, as measured by patient and staff satisfaction, financial performance, and quality outcomes. However, the groups were not able to draw a specific correlation between any one factor and any specific performance metric. This is consistent with the integrated nature of the factors. One of the participants likened the cluster of factors to a seven-atom molecule, in that you could remove one of the atoms and you would still likely have a 'something', but it would not be the original structure. This was a useful metaphor for the group. The groups also affirmed the transitory nature of each of the factors, suggesting the importance of leadership vigilance if the desired agility and resilience were to be sustained.

6 Conclusions

The interviews, literature review, and results of the initial survey are all congruent and, as one reviewer suggested, all seem to align with good leadership practices, which is certainly true. However, I would argue that what has been missing from the discussion about leadership practices in a complex environment has been a coherent, unifying theoretical framework, which I believe complexity theory provides. Further, the survey helps leaders and practitioners move complexity *theory* into something that is far more actionable.

It must be noted that this was an exploratory factor analysis, so caution needs to be shown when looking for effective ways to use the results of subsequent surveys. However, these seven factors have been validated statistically and their presence or absence can now be assessed through an online survey, available through the author. The results of such a survey may help to identify key points of leverage within an organisation. Survey results can then be linked to organisational performance through a series of focus groups, in order to develop new strategies and patterns of engagement.

What emerges from any discussion of factors facilitating emergence is that like the complex adaptive organisations they are intended to influence, no single factor stands alone, but rather it is the interdependence or combinatory nature [24, p. 147] of all of the factors that seems significant.

With this research, practitioners and organisational leaders can take steps to move beyond understanding complexity as a metaphor and begin to see it as a strategic framework within which they can begin to reshape their organisations. However, a final word of caution: this approach shapes and facilitates emergent change, but it is not a 'guarantee' of success. That is simply not in the nature of complexity.

References

[1] Vaill, P.B., Learning as a way of being: Strategies for survival in a world of permanent white water. Jossey-Bass: San Francisco, CA, 1996.

[2] Eoyang, G.H. & Holladay, R.J., Adaptive action: Leveraging uncertainty in your organization. Stanford University Press: Stanford, CA, 2013.

[3] Zimmerman *et al.*, Edgeware; Insights from complexity science for health care leaders. VHA Inc: Irving, TX, 1998.

[4] Plowman, D.A., Solansky, S., Beck, T. E., Baker, L., Kulkarni, M. & Travis, D.V. The role of leadership in emergent, self-organisation. *Leadership Quarterly*, **18(4)**, 341–356, 2007.

[5] Rost, P. *Leadership for the 21st Century.* Praeger Publishing: Westport, CT, p. 102, 1993.

[6] Lichtenstein, B.B. & Plowman, D.A., The leadership of emergence: A complex systems leadership theory of emergence at successive organisational levels. *The Leadership Quarterly*, **20(4)**, 617–630, 2009.

[7] Arena, M.J. Understanding large group intervention processes: A complexity theory perspective. *Organisation Development Journal*, **27(1)**, 49–64, 2009.

[8] Lichtenstein, B.B., Emergence as a process of self-organising - new assumptions and insights from the study of non-linear dynamic systems. *Journal of Organisational Change Management*, **13(6)**, 526–544, 2000.

[9] Olson, E.E. & Eoyang, G.H., *Facilitating Organisation Change: Lessons from Complexity Science.* Jossey-Bass: San Francisco, CA 2001.

[10] Lindberg, C., Nash, S. & Lindberg, C. *On the Edge: Nursing in the Age of Complexity*. Plexus Press: Bordentown, NJ, 2008.

[11] Marion, R. & Uhl-Bien, M. Introduction to the special issue on leadership and complexity. *The Leadership Quarterly*, **18(4)**, 293–296, 2007.

[12] Uhl-Bien, M. & Marion, R. *Complexity Leadership*. IAP: Charlotte, NC, 2008.

[13] Dickens, P., Cultural & global dimensions of change: Power dynamics in a health care setting. Unpublished manuscript, Antioch University, Yellow Springs, OH, 2010.

[14] Dombrowski, C., Kim, J.Y., Desouza, K.C., Braganza, A., Papagari, S., Baloh, P. & Jha, S. Elements of innovative cultures. *Knowledge and Process Management*, **14(3)**, 190–202, 2007.
[15] Heifetz, R., Grashow, A. & Linsky, M., The practice of adaptive leadership: Tools and tactics for changing your organisation and the world. Harvard Business Press: Boston, MA, 2009.
[16] Holman, P., Devane, T. & Cady, S. *The Change Handbook: The Definitive Resource on Today's Best Methods for Engaging Whole Systems*, 2nd edn., Berrett-Koehler: San Francisco, CA, p. 12, 2007.
[17] Wergin, J.F., *Departments that Work: Building and Sustaining Cultures of Excellence in Academic Programs*. Anker: Boston, MA, 2003.
[18] Westley, F., Patton, M.Q. & Zimmerman, B., *Getting to Maybe: How the World is Changed.* Vintage Canada: Toronto, Canada, 2007.
[19] Stacey, R., Complexity and Creativity in Organisation. Berrett-Koehler: San Francisco, CA, p. 72, 1996.
[20] Argyris, C., On organisational learning. Blackwell: Cambridge, MA, 1992.
[21] Mesirow, J., *Transformative Dimensions of Adult Learning*. Jossey-Bass: San Francisco, CA, 1991.
[22] Boisot, M. & Child, J. Organisations as adaptive systems in complex environments: The case in China. *Organisational Science*, **20(3)**, 237–252, p. 238, 1999.
[23] Wheatley, M., *Leadership and the New Science*. Berrett-Koehler: San Francisco, CA, 1994.
[24] Arthur, W.B., *The Nature of Technology: What It is and How It Evolves.* Free Press: New York, NY 2009.

Mutual shaping between technologies and law: memories of Norwegian e-health infrastructures

A. A. Zaytseva
Department of Informatics, University of Oslo, Norway

Abstract

This work suggests a method to study mutual shaping between technologies and law drawing on empirical evidence from two different and simultaneously developing patient-oriented information systems in e-health. The approach of Complex Adaptive Systems is applied to analyze documentary data in the first case to identify causal elements in the 'memory' of its design work, and the narrative data *in situ* and *ex situ* – for subsequent tracing of attractions in the legal environment. A hypothesis on whether both case studies share the same attractions is generated for further testing in the second case. Having replicated the method, the researcher provides explanations regarding the studied phenomenon based on empirical material of the Norwegian e-health sector.
Keywords: attraction, design work, information systems, law, memory, mutual shaping.

1 Introduction

The research is aimed at providing a solution on how to study mutual shaping between technologies and law by use of Complex Adaptive Systems (CAS) approach and narrative analysis. Transdisciplinarity of the research is enhanced by considering design studies of information systems (IS) and the legal research. Two contrasting cases of information technologies within the Norwegian e-health context are selected. One technology is a private solution for storing and sharing health data among health personnel by patients, whereas the second case is a nationwide public sector information infrastructure for critical health data processing in emergency situations. Although these technologies have different legal grounds designwise – one being private and the other public – both are so-

WIT Transactions on State of the Art in Science and Engineering, Vol 90, © 2016 WIT Press
www.witpress.com, ISSN 1755-8336 (on-line)
doi:10.2495/978-1-78466-155-7/005

called 'patient-oriented' technologies, which implies their similarity. These technologies imply high involvement of patients in the decision-making process during their treatment, which is a challenging issue for the Norwegian legislators to ensure. As of 2015, Norway still follows the technology-specific legal approach [1] that makes legislation co-evolve [2] with the emerging technologies tightly and simultaneously.

In the Complex Systems Science has emerged an interest to study peculiarities of the legal realm, e.g. the resilience of the European supranational legislation to changes [3], where networks of legal normative acts of varied statuses in hierarchy have been studied in nodes and links to each other based on quantitative modeling. Playing with nodes enabled the researchers to make conclusions about patterns of responsiveness of the European legislation to emerging social issues. However, the mentioned research is limited only within the legal area of interest and regulative focus, as in the legal discipline. In design studies of IS, which tend to imply qualitative bottom-up methodologies, another problem has been observed, particularly the lack of 'theoretical frameworks for the big picture of the relationship between IS and law', because research in the IS field is basically bounded with organizational focus only [4].

The logic *at the edge* in Complexity Science [5] inspires the will to overcome organizational boundaries of the IS studies and regulatory focus of the legal research for the deeper insight into pure techno-legal dimension *at the edge* of the disciplines and cross-over cases.

2 Empirical setting and methodological preconditions

Since the beginning of the new century, Norway has been concerned with improvements in the e-health sector. Priorities that have been highlighted include adequate medical service to residents, ensuring privacy and human dignity of patients, increasing information flow among the healthcare institutions and teams, and involving the patients in the treatment processes. The legislation in e-health co-evolves with the policies taken and consists of two main regulatory levels – laws and bylaws. The laws become more specialized year by year, but they have norms excluding their power in some rare cases, which private technologies use to regulate their activities in agreements.

The case study, which is used in the given research to generate a method for studying mutual shaping between technologies and law, is a private IT solution, which is hereafter called 'HealthBook'. According to its concept, personal health information is uploaded and archived by its owners – usual individuals – and is shared by them for private purposes with the treating health personnel or anyone else. The technology employs a web-based interface for reading, uploading, and deleting this health information, and a server and a smartphone application that exploits camera for the data upload. The design work on HealthBook began in October 2011 and partially ended in June 2013 with a finished prototype. The design process might proceed in 2015. Specificity of this technology lies in its provision of data sharing from patient to healthcare and not vice versa which is typical for the public e-health infrastructures.

The second case study is a public sector IT solution for collecting vital health data for use by health personnel exercising healthcare cross-over organizational boundaries in situations threatening patient's life. Usual individuals can access their health data via the official health portal. The described information system is hereafter called the National Core Journal. Public discussions to develop such a system started in 2007 and preparation to the design work began simultaneously. The National Core Journal is supposed to be fully realized by 2017 in Oslo. As of 2015, this information system has been implemented as a pilot project in a couple of regions in Norway. The technology has been generally regulated by the Act on Health Registries since June 2012 and by an own bylaw on the National Core Journal since May 2013.

In this research, the design process, or 'design work', is defined as a CAS with 'constrained release of energy' [6] of the involved people making a technologic device from scattered technical elements (e.g. formats, platforms, encryption methods, etc.) [7], which are nested within the national legal environment that has its own elements of formal norms and conventional values (e.g. concrete paragraphs and articles in laws, and legally matter values on technologies inspired of design practices) in dynamic interaction with the former in a particular time. Having been inspired by the school on CAS of the Santa Fe Institute, the researcher assumes that the 'design work' has a 'memory' defined as 'retained parts of the input history' [8]. If the researcher assumes that a CAS of a design work in a case study has an own memory, then it is possible to assume that the design work has a memory bank [9], which is found in shared documentation – reports from the design meetings or with external experts, versions of technical specifications, available e-mail communication, legal proposals and opinion letters, etc. that help people to refer to their earlier talks wherein something particular was discussed.

Generation of the method and its replication has been conducted while analyzing documentary data from the design works on the technologies: 50 documents for the first case study and 120 documents for the second one. From the texts, syntactical elements relevant either to technology or law as their smallest particles have been deduced, but interpretation of the elements is neither broad nor specific. Kinds of people ever involved in the design work are also coded for the visual representations. On the drafted visual representation of the memory for design work in the first case, the researcher investigates visible complex networks of argumentation – 'events'. As a result, only few elements from their total amount are called *causal elements* of the design work due to the fact that they collect most of the links in the stepwise argumentation-based reduction of the 'events'. Although these elements cannot possess the whole contextual richness [10], they indicate topics in the memory, which deserve narrative clarification. Ten interviews have been collected. The narrative acts, or interview data *in situ* from people of the design work and *ex situ* with the external experts also being involved, serve as material for retrieval of additional, specific information from the individual experts. It helps to find possible attractions in law from the identified causal elements. Whether the respective attractions are shared between the cases is subsequently investigated within the method.

3 Analysis, method generation, and stepwise results

In order to draft a visual representation of memory for the first case study – HealthBook – the elements of technology and law have been manually extracted from the available documents relevant to the design work. These documents were originally written in Norwegian (Fig. 1).

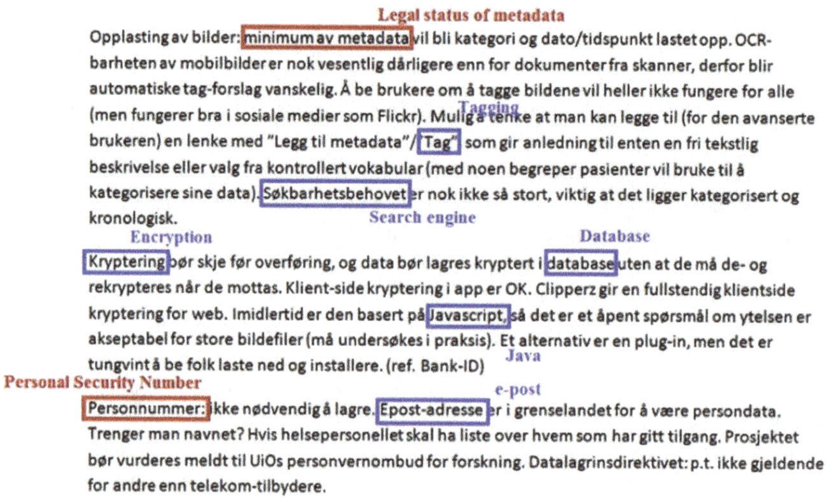

Opplasting av bilder: minimum av metadata vil bli kategori og dato/tidspunkt lastet opp. OCR-barheten av mobilbilder er nok vesentlig dårligere enn for dokumenter fra skanner, derfor blir automatiske tag-forslag vanskelig. Å be brukere om å tagge bildene vil heller ikke fungere for alle (men fungerer bra i sosiale medier som Flickr). Mulig å tenke at man kan legge til (for den avanserte brukeren) en lenke med "Legg til metadata"/"Tag" som gir anledning til enten en fri tekstlig beskrivelse eller valg fra kontrollert vokabular (med noen begreper pasienter vil bruke til å kategorisere sine data). Søkbarhetsbehovet er nok ikke så stort, viktig at det ligger kategorisert og kronologisk.

Kryptering bør skje før overføring, og data bør lagres kryptert i database uten at de må de- og rekrypteres når de mottas. Klient-side kryptering i app er OK. Clipperz gir en fullstendig klientside kryptering for web. Imidlertid er den basert på Javascript, så det er et åpent spørsmål om ytelsen er akseptabel for store bildefiler (må undersøkes i praksis). Et alternativ er en plug-in, men det er tungvint å be folk laste ned og installere. (ref. Bank-ID)

Personnummer: ikke nødvendig å lagre. Epost-adresse er i grenselandet for å være persondata. Trenger man navnet? Hvis helsepersonellet skal ha liste over hvem som har gitt tilgang. Prosjektet bør vurderes meldt til UiOs personvernombud for forskning. Datalagrinsdirektivet: p.t. ikke gjeldende for andre enn telekom-tilbydere.

Figure 1: Extracting elements of memory from the documentary data.

As a result, 51 technological elements were collected and placed at the top sector of the memory, 16 elements of persons in the design work were placed in the middle sector, and 23 legal elements were placed at the bottom sector (Fig. 2).

The earliest mentioned phrase with the element provides it with a time-tag of when it was introduced for the first time. Having been mentioned once in the text, an element gets its own 'track' in the memory and subsequent time-tags corresponding to the dates of the documents when the element appears. Authorship of the element is designated with a color. When mentioned again by any other person of the design work besides the original author, the element re-appears in the track and gets colored correspondingly. Afterward, links are built among the elements based on authentic argumentation as in the text: naming one after another as a suggestion, introducing one in the form of question or in negation, etc. Visual representations are handy for illustration of complex phenomena [11]. So, in Fig. 2, we notice seven complex networks, or 'events', with elements having their node degrees of three or more. These seven complex networks deserve a closer look. They are highlighted in Fig. 3.

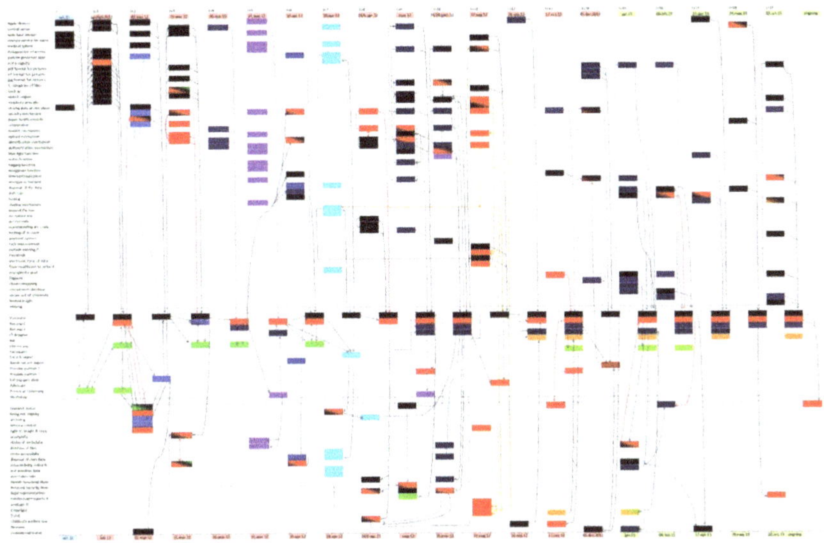

Figure 2: Memory of design work in the first case study.

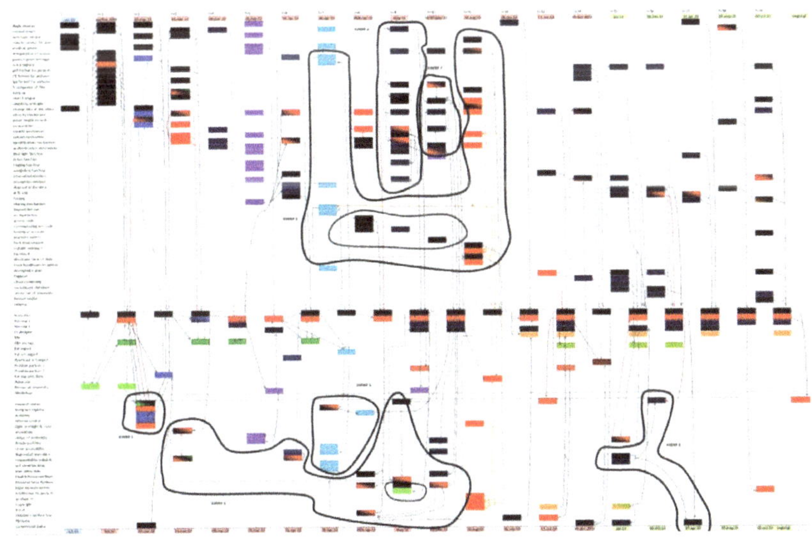

Figure 3: 'Events' in the memory of design work in the first case study.

Thus, from the visual representation of the memory with 'events' (Fig. 3), the researcher retrieves elements composing each 'event' (Fig. 4). For clarity, the elements that belong to specific 'events' have been colored to indicate where they have been extracted from in Fig. 3.

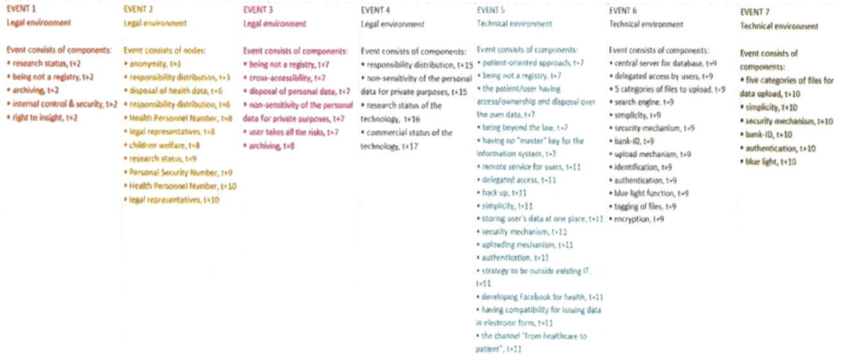

Figure 4: Elements of 'events' in the memory of design work in the first case study.

We may notice that time-tags of the elements inside single 'events' vary, as in the 'events' numbered 2, 3, 4, and 5. For the next visual representation, the elements of 'events' are reorganized in a sequential order of their time-tags. At this step, the researcher drowns on her own memory. It is required to remember empirical data, so that drafting of relations among the elements in the visual representation forms a meaningful storyline consisting of those elements that are found in close proximity. While doing this, we shall keep in mind that building a meaningful story from the available elements implies finding logic first within one 'event' in different time-tags, and then, when the logic seems to fade, among the elements of different 'events' within the same time-tags (Fig. 5).

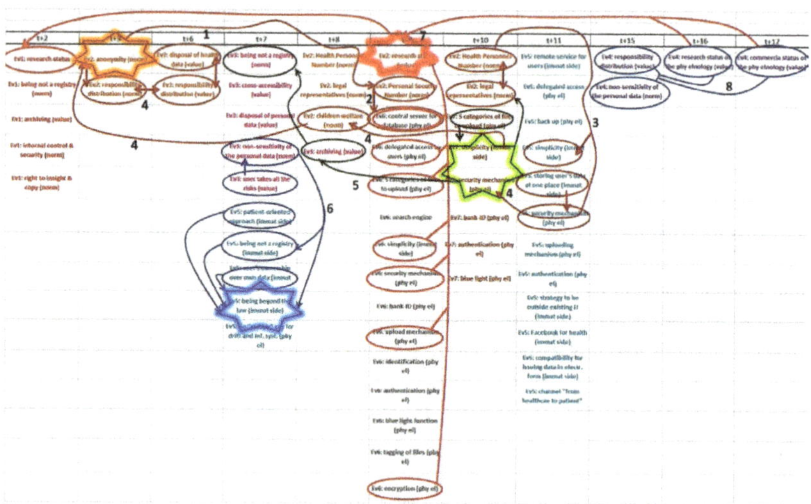

Figure 5: Tracing causal elements among the elements of 'events' in the first case.

For example, if for the 'disposal of health data' the issue of 'redistribution of responsibility' among the users of HealthBook was raised, then 'anonymity' of the users that is possible within the concept of patient-ownership of health data emerged as opposite (1), but not contradictory to identification. Identification tends to include in Norway 'Personal Security Number' implying 'central server'; it stores the patient data accessible via identification mechanisms (2); authentication that is secondary to identification happens via 'Health Personnel Number', a property of 'security mechanism' adapted specifically to HealthBook for its 'archiving' purpose, which puts the technology into a possible legal status of 'health registry', something that the technology concept wants to avoid (3). The same logic is applied further.

In this multi-step, non-linear qualitative analysis, the elements collecting more than three arrows are marked as stars – they are *causal elements*. These are 'anonymity', 'simplicity and 'security' considered together, 'being beyond the law' and 'research status' of HealthBook. The use of interview data as narrative acts helps the researcher to retrieve further details regarding key reasoning and concerns [12, 13] in the decisions taken on technologic implications of these elements in the design work. Since the professional meanings of the people vary depending on their experience and impression from the design work, the researcher is able to retrieve rich contextual information [10] to make conclusions about the attraction in the legal environment to the final decision-making via the causal elements as the most sensitive in the design work.

From the narrative data, it was found that the technologic implications to 'anonymity' were stating username and e-mail address of the individual for registration and use of HealthBook that does not have any technologic channel of verification to the user's identity, when the identity could only be disclosed from the shared health data itself. The legal side of this is that law does not impose any duty to design verification mechanisms if the technology has such architectural constitution as HealthBook – the cloud computing model – and, therefore, as the involved *in situ* designer says, 'the system does not discriminate' its users. This technologic detail is, however, unclear to the users when they create an account and accept the terms and conditions, which raises a legal attraction or concern [12, 13] about how much the user shall be informed about the technology.

The second causal element 'simplicity and security' has technical implications in its sharing mechanism that is simple, but implies stating health personnel number, encryption methods, and entrance code generation. Legally speaking, this is not the highest level of security that is supposed to be applied in the Norwegian e-health sector, but HealthBook has a unique security framework and informs the user in the terms and conditions that s/he takes all risks, which is considered as an agreement between the individual and the e-health service provider. This fact raises a legal attraction on the responsibilities of the e-health service provider – do they exist and what are they built upon?

The third causal element is 'being beyond the law', which from the narrative material determines the lack of own database for storing health data in the architecture of HealthBook. This corresponds legally to the lack of duty to internal control and to the architecture not being considered as a health registry. Not being a health registry is indirectly allowed by par.2 §3 of the Act on Personal Data, which excludes the power of this law and other related laws in case sharing of sensitive personal data takes place for private purposes. The database, which lacks in architecture, indicates a legal attraction relating to the issue on how to define a data controller – one who is responsible for data processing.

The fourth causal element 'research status' determines the technical implication in the form of a server within the cloud computing model of HealthBook with the university possessing it. In legal terms, HealthBook is a research project, whereas in accordance with its concept, it was supposed to be a novel, privately owned 'patient-governed journal' with commercial implementations. 'Research status' raises an attraction on what legal nature patient journal systems may have and what kind of statuses they might balance in between.

Since both of the studied IS exist in the same legal environment – the national legislation of Norway – the researcher may assume that their design works might share attractions. This forms a hypothesis for subsequent steps in the method. However, due to the differences of the cases in their scale and time consumption for design works, the causal elements might unforeseeably differ in forms/shapes as expressed in the documentary data.

In order to test the hypothesis, the same method of finding causal elements and areas of attraction is applied to the second case study, which is a nationwide e-health infrastructure for critical health information. The rules of drawing the memory for the National Core Journal (Fig. 6) are identical to those formulated concerning the memory of HealthBook.

The only difference is that there is no need to interrelate the technology and law sectors with the sector of actors involved in the design work, because exchange of professional opinions in documentaries took place not at the shared design meetings, but via communication of official documents by deadlines as replies to the previously published reports and legislative proposals. Besides, the 'events' of visual complexities are only to be found in the sectors of technology (on the top) and law (on the bottom) in separation, as it was done for Fig. 3. Consequently, their elements are to be interconnected in the subsequent step of the story-building by the researcher in accordance with the rules prescribed for Fig. 5. That step would reduce 'noise' of multiple voices from 77 people/agencies in the design work on the National Core Journal toward 451 elements of technology and 346 elements of law.

In the provided visual representation of memory in the second case study, we can visually notice 14 complexities, or 'events', for further non-linear analysis. Due to limitations of the page format for writing this article, the figure with the marked 'events' is not provided, but it is necessary to hint that the 'events' have horizontal positioning and are highly interconnected with each other. The researcher would notice that while unraveling their elements via links and organizing them in columns for the second case study in the same way as it was done for the first case study (Fig. 4). The rule to unravel the 'events' is based on

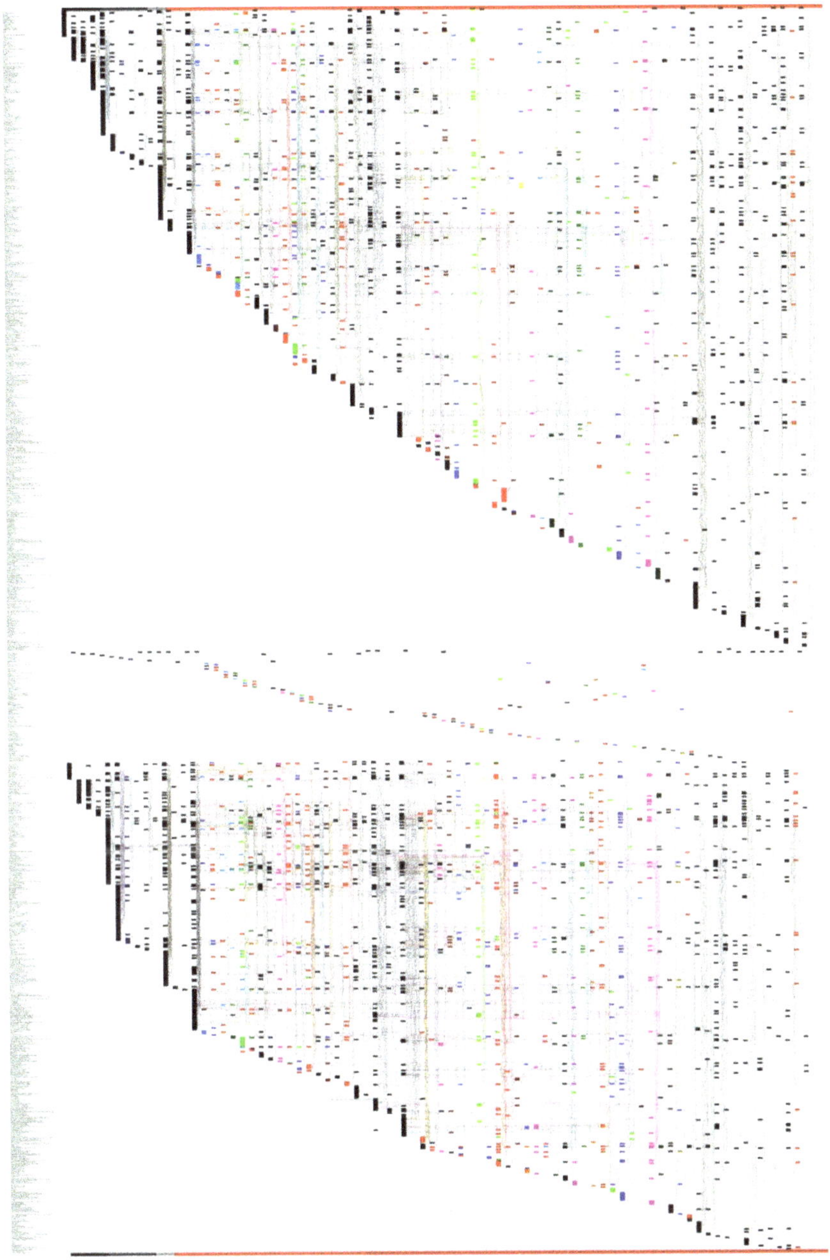

Figure 6: Memory of design work in the second case study.

tracing elements with a link of a color other than their own. If an element belongs to more than one 'event' due to its high interconnectivity with the other 'events', then the element is considered belonging to the 'event' where it was discovered first. The visual representation for history-building and finding the causal elements in the second case is shown in Fig. 7.

According to the story-building conducted among the elements, as based on the rules to Fig. 5, Fig. 7 illustrates that the causal elements in the design work on the National Core Journal are 'patient consent', 'centralization of information', 'synchronization', 'control over data by patient' (met two times), 'sensitivity of health data' (met two times), and 'automatic update'.

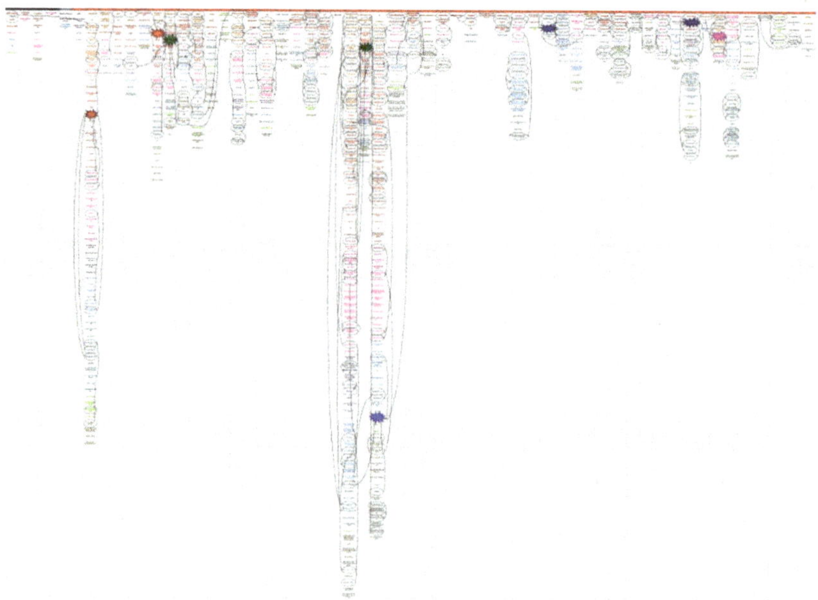

Figure 7: Tracing causal elements among the elements of 'events' in the second case

In terms of 'patient consent', its technological implication is the mechanism of access to health data in the National Core Journal designed to allow insight into it on having registered a consent given to health personnel by a patient. However, as it comes from the narrative data with the leading *in situ* designer of this technology, to have patient consent as the only legal ground 'would make the National Core Journal impossible' as in all countries where access to e-health infrastructures is based on patient consent. Therefore, the legal side of the access to the National Core Journal is also the right of patient to reserve him/herself from this IT solution, what would mean risks of wrong healthcare provision, injuries, or death in case of unplanned hospitalization. Before using the right to reserve against the National Core Journal, the patient shall be completely informed about the full set of services that this information infrastructure provides. To be informed about the right of reservation against the National Core Journal means also being informed about consent giving. As the designer claims, having the user informed is a difficult legal issue in case of e-health service provision based on agreements with private e-service providers. This finding is shared with the attraction for 'anonymity' in HealthBook.

'Centralization of information' as a causal element implies storing vital health data at one place for sharing cross-over health organizations. As the *ex situ* expert of the design work mentions in the narrative empirics, the National Core Journal became 'an alternative solution toward electronic cards of health personnel sharing patient data for treatment purposes'. Legally speaking, this intention led to creation of the norm §6d in the renewed Act on Health Registries and migration of its §13 on data sharing only within one organization to the new §19 on cross-organizational health data sharing in the Act on Patient Journals. Although the National Core Journal is named so, it is not a conventional patient journal system; it is 'an addition to it' and a health registry, but not under the main regulation by the Act of Health Registries. Balancing between two legal realms make the National Core Journal share the attraction on the unclear nature of patient journal systems, as found via 'research status' of HealthBook.

Although the National Core Journal creates an impression of an integrated infrastructure, it is not so. From the narrative data *in situ* it follows that serious difficulties have been experienced in the technical integration of IS for the National Core Journal. Electronic patient journals have varied specifications and are delivered by competing providers in Norway, whereas the public-sector Health Personnel Registry and the Population Registry have not been designed for further integration at all. 'Synchronization' as a causal element in the design work implies elaboration of the common requirements to technical functionalities and redistribution of legal responsibilities. The interview data reveals that defining data controllers in the relevant models of infrastructures in the synchronization process is difficult. Thus, the legal attraction on data controllers, as indicated by 'being beyond the law' in the first case, is shared with the design work in the second case as well.

The causal element 'sensitivity of health data' has its technologic implication for hardware in architecture, whereas legally it would relate to the ownership of this equipment and the right to data disposal contained. If the data are stored on a private device of the patient, s/he owns this data and has the right to dispose it for any purpose. If the data are stored on equipment that belongs to anyone else, then the processing of this data – storage and sharing – shall be exercised by the data processor only toward providing proper healthcare to the data owner as guaranteed by the legislation on healthcare. 'Taking measures for proper security mechanisms corresponds the duty of proper healthcare provision', claims an *ex situ* expert in the design work of the National Core Journal. Thus, the attraction on duties of e-health service providers, as indicated for HealthBook, is not necessarily shared with the National Core Journal case that does not question own duties to the highest level of security. However, this attraction can be shared with another public sector e-health technology – the Personal Health Archive – that is designed for offering services for storing private health data.

The two remaining causal elements 'control over data by patient' and 'automatic update' ground a possible bridge of attractions with other design works on technologies that can be studied further with the method replicated to any third case in the legal environment. 'Control over data by patient' has its technological implications in logging functionality and numerous notifications to the patient on data processing. Legally this element implies the right of patient to block own health data against some or all health personnel groups, or block some or all health records, which may indicate the attraction of emerging patient rights to govern information elements. The last, but not the least, 'automatic update' implies collection of critical health data conducted fast and in a structured form which would ensure neutrality of health data interpretation and decreasing probability of fatal medical errors. The possible legal attraction can be relevant to study sanctions on health personnel and responsibility taking for professional errors while using information infrastructures for healthcare provision.

4 Conclusions and further research

The research presents a method for stepwise qualitative analysis and explanation of the phenomenon of mutual shaping between technologies and law, irrespective of the number of cases and their 'appropriateness'. The method contains not only extraction of wordings from qualitative data, but also meaning-based analytical steps, where narrative analysis is applied for building sets of a story on simultaneous design works within a shared legal environment. If the CAS approach forms a basis for pattern recognition and causality tracing, then the narrative approach contributes to universality of the generated method. This work addresses emerging discussions in Complexity Science on the importance of bottom-up techniques used together with algorithmic, stepwise methods of the highly logic systems thinking. The method generated in this research provides a solution to investigating the wholeness of a space *at the edge* of the legal sphere of competence and techno-organizational expertise, and reduces the 'noise' of their multiple subjective points of view about what can influence the technology and how the technical solution in its original shape might respond to the law. Furthermore, the method may help analysts to anticipate environmental conditions enabling or constraining technologies that are in the early stages of design work, which might help to prevent cost overruns. Application of the method to empirical, real-world data indicates legal attractors as pointers to relevant legal instabilities that influence strategic decision-making regarding unique combination and modularization of technological elements in unity with ongoing techno-legal co-evolution.

References

[1] Aalberts, B.P. & van der Hof, S., *Digital signature blindness: analysis of legislative approaches toward electronic authentication*, Tilburg, November 1999, online publication, http://heinonline.org/, pp. 2–72, 1999.

[2] Rzevski, G. & Skobelev, P. *Managing Complexity*, WIT Press: Ashurst, 2014.

[3] Koniaris, M., Anagnostopoulos I. & Vassiliou, Y., Network Analysis in the Legal Domain: A complex model for European Union legal sources. *Physics and Society*, Cornell University Library. Online publication, http://arxiv.org/abs/1501.05237

[4] Knackstedt, R., Eggert, M., Heddier, M., Chasin, F., & Becker, J., The relationship of IS and law: the perspective of and implications for IS research. *ECIS 2013 Proceedings*, paper 18, 2013.

[5] Waldrop, M.M., *Complexity: the Emerging Science at the Edge of Order and Chaos*. Simon & Schuster Paperbacks: New York, 1992.

[6] Kauffman, S.A. *Investigations*, Oxford University Press: New York, 2000.

[7] Feng, P. & Feenberg, A., Thinking about design: critical theory of technology and the design process. *Philosophy and Design: From Engineering to Architecture,* ed. P.E. Vermaas, P. Kroes, A. Light, & S.A. Moore, Springer Science Business Media B.V. pp. 105–118, 2008.

[8] Holland, J.H., *Adaptation in Natural and Artificial Systems: an Introductory Analysis with Applications to Biology, Control, and Artificial Intelligence*, Massachusetts Institute of Technology: Massachusetts, 1992.

[9] Wegner, D.M., A computer network model of human transactive memory. *Social Cognition*, **13(3)**, pp. 319–339, 1995.

[10] Tsoukas, H. & Hatch, M.J., Complex thinking, complex practice: the case for a narrative approach to organizational complexity. *Human Relations*, **54**, pp. 979–1013, 2001.

[11] Kirsh, D. Thinking with external representations, *AI & Society*, **25**, pp. 441–454, 2010.

[12] Luhman, J.T. & Boje, D.M., What is complexity science? a possible answer from narrative research. *Emergence*, **3(1)**, pp. 158–167, 2001.

[13] Byrne, D. & Callaghan, G. *Complexity theory and the Social sciences: the state of the art*, Routledge: London, New York, 2014.

'In' or 'as' space?: a model of complexity, with philosophical, simulatory, and empirical ramifications

C. H. Smith
University Libraries, Western Kentucky University, USA

Abstract

A General Systems model based on ideas originating with the writings of Benedict de Spinoza is described, starting with its philosophical underpinnings and proceeding on to its relation to modern systems concepts, including attempts to simulate the relationships posed and measure real-world structures. Central to the idea is the notion that spatial extension may not have a prior existence but emerges only through an entropy maximization process in which information and energy exchange is balanced among some limited number of subsystems that in sum comprise any given functioning complex system. Related published empiricism concerning geographical/geological systems – the hypsometry of stream basins and overall internal zonation properties of earth structure – is briefly described; the former, especially, reveals a hierarchical pattern of potential energy relations that seems to fit well the organizational hypothesis. Possible applications of the model to genomic codon and medical imaging modeling are alluded to. A brief treatment of the relation of the model to basic properties of complex systems (connectivity, autonomy, emergence, etc.) is provided.
Keywords: complex systems, entropy maximization, General Systems Theory, simulation, space, spatial extension, spatial structure, Spinoza, stream basins.

1 Introduction

Several years ago, Wilson noted that 'What characterizes systems of organized complexity is essentially that they are made up of large numbers of parts – and that these parts are strongly connected; that is, they each interact strongly with a number of others' [1, p. 634]. In a follow-up paper, he went on to suggest that

WIT Transactions on State of the Art in Science and Engineering, Vol 90, © 2016 WIT Press
www.witpress.com, ISSN 1755-8336 (on-line)
doi:10.2495/978-1-78466-155-7/006

Boltzmann methods might be used to make 'spatial interaction terms explicit', thereby giving 'a new ecological model' that could provide 'a potentially more effective way of handling spatial interaction', and 'ways of analyzing the dynamics of spatial patterns' [2, p. 865]. In this work, he mentions that 'There is only one empirical example in the literature known to the author', citing an early paper by the present writer [3]. Here I would like to present the outlines of a model relatable to these thoughts, but emerging from a surprisingly different starting point.

As a graduate student in the late 1970s and early 1980s, I took a number of courses in the area of history and philosophy of science. One of these focused on the Vienna Circle school of logical positivism, and especially a famous work by Carnap, *Der Aufbau* [4]. This attempt to generate a logical positivism underpinning for all of science ultimately fell through when Quine [5] observed that Carnap's system of associating, and relating, objects in space neglected to provide a satisfactory definition for the elemental concept 'located at'. I began to search for a different way by looking at the concept of interaction in space, and found what I was looking for a couple of years later when I sat in on a class featuring a review of rationalist philosophers, including Benedict de Spinoza (1632–1677).

Spinoza's assessment of the structure of natural reality, as expressed in his *Ethics*, involved a rather different starting point from the Empiricist approach of Locke and others. Spinoza's reasoning out of the nature of existence (backed by transcendental proofs that no one seems to understand fully) begins with the posing of what he terms 'Substance'. 'Substance' is removed from any kind of direct access by ordinary mortals, seemingly representing the very essence of existence (God? or some pantheistic conception thereof?). 'Substance' expresses itself through vehicles that Spinoza terms 'Attributes', though these are not the directly accessible/measurable attributes of the Empiricist tradition. Instead, they seem to represent some kind of 'rules of order' through which the mundane, experienceable, elements of existence (which Spinoza refers to as 'modifications' and 'modes') are able to express themselves. Spinoza argues that humankind is capable of fathoming only two basic 'Attributes': thought and spatial extension. Importantly, each of these independently can provide a complete appreciation of the mundane elements of existence, and in the same basic manner for all of the latter. Recall the old concept of 'all of nature in a rose'.

It quickly occurred to me that perhaps the secret to such 'rules of order' lay at the level of subsystem interactions within each discrete natural system, that is, that perhaps one could divine some basic property or properties of internal interaction (material, energy, and/or information transfer) fundamental to the organization of all such entities. As a first attempt, I set out a model in which natural systems were viewed as subsystemizing according to a 'lazy universe' manner of hierarchical organization [6]. This approach was based on an analysis of hierarchical inclusion patterns involving combinatorial mathematics, as shown in Fig. 1. There, one can view how a hierarchy of relations beginning with 'first-order' subsystems might self-organize. A model of one of the two basic

"Attributes" of Spinoza was thus achieved (this is probably best understood as a model of the Attribute "thought", though I did not recognize this at first).

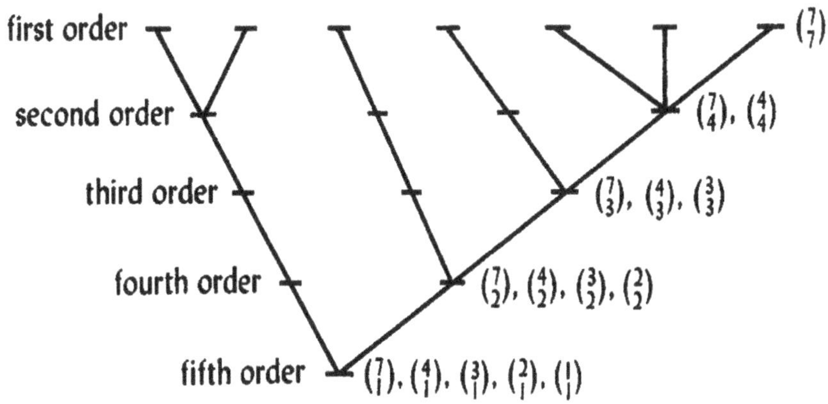

Figure 1: A 'maximum likelihood tree' of hierarchical relations constructed with the aid of combinatorial mathematics. Within this structure, all class–subclass relations are ordered on a 'most-probable-state' basis. Thus, and for example, the seven first-order subclasses group most probably into one second-order class containing three first-order subclasses, one containing two subclasses, and two containing one subclass. All other relations noted observe the same basic plan. When the structure is extended beyond seven initial elements, it begins to lose this property. (After [6], Fig. 5.)

It is probably possible to design empirical tests bearing on the validity of this model, but as it represents an understanding of historical differentiation, and thus in practice one-of-a-kind evolutionary processes, it seemed preferable to try to think out a model more amenable to the ecological realities of spatial extension (the other Spinozian attribute). What emerged was a somewhat audacious structure, one which addresses the matter head-on by posing that, perhaps, spatial extension itself represents an entropy maximization process originating with the characteristics of interaction among a given natural system's (each corresponding to a Spinozian modification) ecological subsystems.

Perhaps the easiest way to imagine this is to think in terms of an $i = j$ – dimensioned input–output matrix depicting a state of standing interaction. A familiar geographical example is the standing pattern of commuter trips among all combinations of towns in some metropolitan region. Assessment of the relative rates of such flows is complicated by the varying populations of the towns involved, but the earlier-mentioned Wilson got past this difficulty by applying entropy maximization methods to the problem [7]. A 'most-probable-state' solution to the matrix of interactions was thus achieved which accounted for a large proportion of the variation present in the original flows matrix; once this effect was removed, further causalities (often related to the relative distances among the places involved) could be investigated.

This approach is still in use. To a certain extent, it has lost its appeal as a modeling device, however, devolving into a methodological tool with apparent limited relevance to how urban/regional structures actually *evolve*. The problem, it seems to me, lies in restricting the measurement of flows to arbitrarily defined sources and destinations that do not necessarily reflect the manner of organizational differentiation within natural systems. The papers of Wilson [1] and Wilson *et al.* [2] address this issue.

If we could identify a form of subsystemization that is basic to the structure involved, and whose operation can be expressed in the form of measurable interactions among the elements, we might be better able to get at the relation of system change to its ongoing maintenance of ecological integrity. I conceived of a system structure characterized by a small number of subsystems: (1) that in sum constitute the whole of the system; (2) that share information, energy, and/or materials with one another; (3) in a manner that can be objectively represented as 'flows' (implied or actual) in an input–output matrix. Finally, and most interestingly, I think, I propose that an entropy maximization of this kind of intra-system structure might reveal a pattern of symmetry interpretable as three-dimensional space. Otherwise put, I introduce the possibility that it is a continuing entropy maximization of this kind of relations that *is* spatial extension, to the extent that everything that is spatially extended must observe these conditions.

This solution, if it be a real one, is one which seemingly remains faithful to the Spinozian model, while lending itself to examination as something more than just a philosophical structure. In particular, it appears to be particularly amenable to simulation and, indeed, to actual empirical investigations.

2 Simulations

In attempting to simulate such a set of relations, the most direct approach involves brute force methods. Basically, are there any sets of numbers, expressed in dimensionality $i = j$ matrix form, that when entropy-maximized will produce the anticipated results? If there are none, of course, the thought process here is for naught; conversely, if all sets tested yield the necessary outcome, the model produces, at best, trivial results.

The specific question being investigated in the simulations is whether a given matrix of numbers, when entropy-maximized, will produce results interpretable as a Euclidean space – that is, which can be projected as a set of Euclidean dimensional relations. To examine this, I filled hundreds of thousands of matrices, of dimensions 3×3, 4×4, 5×5, and 6×6, with six-digit random numbers, and submitted these to an entropy maximization operation known as double-standardization (or bistochastization), in which the rows and columns of the matrix are alternately converted to z-scores for as many times as it takes to converge on stable values. It turns out that only one kind of result directly corresponds to (i.e. 'produces a projection of') a set of relations recognizable as

Euclidean spatial extension: for a value of $n = 4$ matrix dimensions, $ij = ji$ scores that are symmetric about the main diagonal. (This can be confirmed through metric multidimensional scaling, the result being a zero-stress solution wherein all the unrotated three-dimensional coordinate points are effectively indistinguishable from one another.) For example, this matrix of raw numbers:

2835	1962	1293	908
1962	2287	2098	1689
1293	2098	2500	2442
908	1689	2442	3615

when double-standardized, produces the z-scores:

1.4562	0.2296	−1.2934	−0.3924
0.2296	1.4562	−0.3924	−1.2934
−1.2934	−0.3924	1.4562	0.2296
−0.3924	−1.2934	0.2296	1.4562

These scores, when interpreted as similarities and transformed into distances, correspond to relations within an unambiguously defined three-dimensional space.

Depending on the exact initial conditions imposed on the simulations, the percentage of $n = 4$ matrices that achieve this symmetric form ranges between about one-half of 1% and 2%. Figure 2 gives a summary of the most important of these results, combining several pieces of information. For each matrix dimensionality tested (again, 3×3 through 6×6), four sets of simulations were run, combining two different conditions of random number placement/inclusion (fully random, or random but symmetric about the $i = j$ diagonal) with two secondary statistical measures of the inputted data (mean value of the r values from the associated Pearson r correlation matrix or mean value of the absolute values of the column means of same). Points 17 through 20 in Fig. 2 represent the parallel data for z-score results that passed the $i = j$ symmetry requirement just described; again, these occurred under the $n = 4$ dimensions condition (and thus represent a special subset of the data leading to points 2, 6, 10, and 14). The standard deviations associated with the 20 mean values plotted are rather large, but it is apparent that, broadly speaking, the mean r values decrease with increasing numbers of classes. The significance of this fact will become apparent in a moment.

The results displayed in Fig. 2 make sense in several crucial respects. First, the fact that 17, 18, 19, and 20 exist at all signifies that a small, but not trivially small, number of appropriate solutions exist under the universe of all possible matrix contents (i.e. simulated interaction states). Second, these findings are sympathetic to the fact that four nodes are required to define a three-dimensional Euclidean space. Third, the high standard deviations connected with each mean plotted as 17, 18, 19, and 20 above indicate a wide range of intra-correlations among the vectors of the matrices involved, a potentially important clue to the way systems may evolve internally toward reduction of redundancies of function through feedbacks. Fourth, and a related consideration, equations representing the functions of systems involving five or more variables are almost never

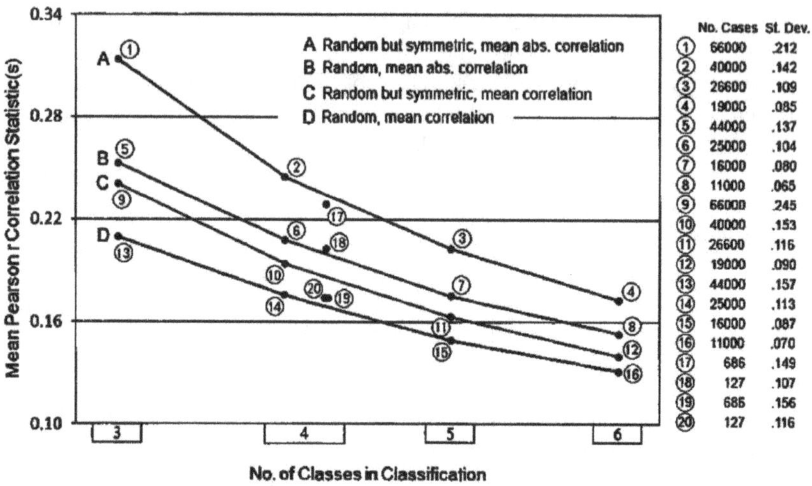

Figure 2: Summary of spatial projection simulations for random numbers matrix configurations of dimension 3 × 3 through 6 × 6. Circled numbers refer to data at the right margin giving the number of simulations in each test and the standard deviations accompanying the mean values plotted. The plotted values are the means obtained for each set of simulations performed at each matrix dimensionality (='no. of classes in classification'). Lines connecting the points are for readability purposes only. Point values 17 through 20 are compiled from subsets of the data leading to point values 2, 6, 10, and 14, respectively. See text for further explanation. (After [8], Fig. 1.)

solvable, making it possible to imagine how a four-subsystem structure might self-adjust.

The success of the random numbers simulations suggested an elaboration. So, soon after completing them, I began to investigate point patterns (spaced regularly on geometric surfaces) grouped randomly, or arbitrarily, into class structures (here, of four classes). In this instance, the summary values filling the raw data matrices were metric spatial autocorrelation scores (which provided a measured indication of the relative overall location of the members of each class of 'sampled' points with respect to the overall locations of the members of each other class). These studies are described elsewhere [8–10]. Briefly, I found that roughly 5% of the systems spreading out over arbitrarily bounded two-dimensional surfaces passed the symmetrical z-scores test. This percentage increased further when several configurations within, and across the surface of, spherical systems were analyzed. In all, an 'evolutionary' sequence is suggested in which it becomes easier and easier to achieve a symmetric z-scores result, as the prior geometry changes from undefined, to two-dimensional, to three-dimensional.

3 Real-world systems

The success of the simulations provided an impetus to continue on to empirical analyses of actual natural systems. A number of pilot studies on systems ranging from butterfly wing color patterns to regional human populations were attempted, with some success [8–10]. Here I would like to describe the two most successful of these, which produced published works [8, 9].

In this context, it would be ideal to investigate systems characterized by actual and discrete energy, materials, or information flows that can be directly measured, but the body of knowledge that could sustain such measurements may not exist as yet. The next best thing seems to be an investigation of patterns which represent 'implied flows', that is, the structural result, in space, of continuing processes. For good examples of these, we may turn to topographical and geological models.

Stream basins represent bounded systems characterized by relatively well-worked out topographical patterns that are due largely to a simple universal driving mechanism: gravity. Thus, every point on the surface of any given basin has a certain potential energy connected with it, and ultimately it is the working out of these potentials with respect to one another that leaves us with the topography witnessed at any given time. The question posed here is whether such interactions operate as singularities (as do the commuter trips between a large number of origins and destinations) or actually respond to causal forces operating primarily at a grouped level (most likely, as ranges of elevations).

As described in more detail in the published account, the elevations within 31 similarly-sized drainage basins in the Commonwealth of Kentucky, USA, were sampled at three levels of spatial fineness to produce three sets of data; each of these were then assembled into rank-ordered vectors. These data were then subjected to a series of nonhierarchical, information statistic-based, cluster analyses, classifying the contents of the 93 vectors into 3-, 4-, 5- and 6-class solutions.

The cluster analyses produced unexceptional results diagrammed as Fig. 2 of [8]; at all three fineness levels of sampling, the variations explained increased, on the average, in a very regular pattern according to the number of classes imposed. But this analysis did not take into account the relative spatial positions of all the point elevations sampled, and when this was done, a very different picture emerged.

All 93×4 sets of data were subjected to a pair of metric spatial autocorrelation analyses, that is, the spatial relations of the sampled points within each range of elevations were related, as a group, to the relative spatial positions of the points within the other groups (and to themselves). This produced $93 \times 4 \times 2$ matrices of spatial autocorrelation coefficients (of dimensions 3×3, 4×4, 5×5, and 6×6), which were then double-standardized as described in the simulations section.

Of the results, as was expected, none of the operations on the three-, five-, and six-class structures produced the symmetric z-scores pattern described earlier. For the four-class solutions, 28 of 31 basins passed the test at the finest sampling levels, but only 18 of 31 did at the coarsest one (the real-world distance between nearest neighbors in the finest sampling grid was 215 feet, the second finest 430 feet, and the coarsest 860 feet). A special, even finer, grid was applied to sample the three basins that did not pass the test, whereupon two of the three then passed the test, with the third showing considerable movement toward passing, but not quite doing so.

A perhaps even more interesting result is depicted in Fig. 3. This chart is organized in a manner similar to Fig. 2, but instead of displaying the results of simulation exercises, it summarizes the stream basins analysis. All of the values plotted represent means across each set of 31 results (the numbers next to each datum point are the accompanying standard deviations). Two sets of results are presented, based on the two internal redundancy measures employed. Within each set, the results for each classification model at the three different finenesses of sampling are shown.

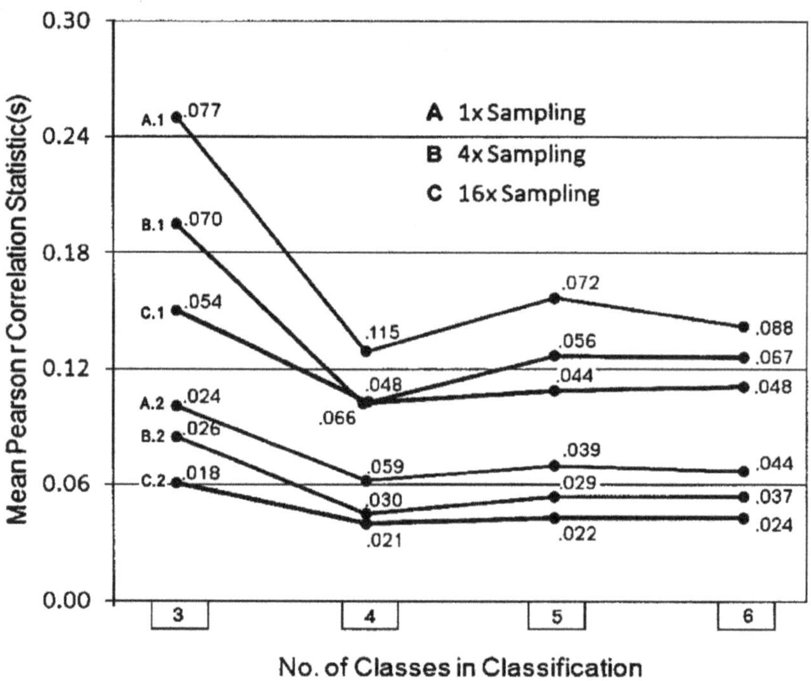

Figure 3: Summary of spatial subsystemization properties of topography in 31 Kentucky stream basins, based on three fineness levels of sampling and applying the two mean correlation matrix formulations (top set of three, mean of the absolute values; bottom set of three, mean of the means). The associated standard deviation for each mean value plotted is written out next to it. The lines connecting the means plotted are for readability purposes only. See text for further explanation. (After [8], Fig. 3.)

In contrast with Fig. 2, the trend toward decreasing mean r values with increasing number of class divisions is broken up abruptly at the four-class level; this is apparent in all six sets of models and is thus a very robust result. In fact, in all six sets of models, the *lowest* mean r value is associated with the four-class model. It thus seems apparent that *something* is producing a special causal effect at this level.

Again, the significance of the mean r statistic is that it possibly provides a glimpse into the organization of redundancies within a system – at least to those operating at the subsystem level as defined here. The lowest mean r possible for a correlation coefficient matrix is zero; under such conditions, each vector of values in the matrix is as different as is mathematically possible from the other - vectors, taken as a group. It is tempting in the present instance to regard real-world examples exhibiting rather high mean r scores as representing systems that are unstable, that is, as operating under the influence of feedbacks that are serving to move the entity toward higher levels of efficient operation. The results of the analysis discussed here seem to fortify this impression, as the three-, five-, and six-class models show no evidence of any trend beyond that of a simple statistical nature (similar to the increasing 'variation explained' cluster analysis results discussed earlier), whereas the four-class model does.

One may reasonably ask whether the results displayed here might represent simple reconstructions of the three-dimensionality already built into the setting, but three arguments can be presented to show that this is an unlikely interpretation. First, the mean r values for the four-class solutions here are much lower than for the parallel simulations noted earlier. Second, a follow-up analysis was performed on the 1× and 16× samplings in which the vectors of elevations were arranged into unnatural (but still range-of-values-related) class structures; when this was done and the resulting spatial autocorrelation statistic-based matrices were entropy-maximized, only a small percentage of the models passed the z-scores symmetry test, and the resulting mean r values approached those from the simulations. Third, in an earlier pilot study involving 25 stream basins, results consistent with (but not so striking as, being based on coarser data) the ones discussed here were obtained, while a different kind of follow-up analysis was administered. It was reasoned that those individual basins that yielded high mean r values should be associated with topographical basin conditions indicative of imbalances between internal rates of erosion and deposition: in short, that the slopes across such basins should be less regular. A fairly crude measure of same was applied, with the result that nearly 50% of the variation among the 25 mean r values was explained by this variable alone [10].

The second analysis, published in 2014 [9], may now be described briefly. It also involved a spatial autocorrelation-based approach. Many natural systems, as interpreted through the model discussed here, might be expected to exhibit rather high internal redundancies, either as a result of natural development trends or in temporary response to external influences (as would be the case with the stream systems noted above). I wanted to investigate some system that one would expect to have highly settled internal relations, the results of a very long and relatively undisturbed evolution. The internal zones of the earth seemed to fit this description; so, I have undertaken several analyses of the spatial relations among the inner core, outer core, mantle, and 'surface zones' (oceanic and continental crust, plus hydrosphere and atmosphere). The results of these studies may be found in [9, 10]. Earlier simulation studies [10] on the development of concentric zones with spheres and spheroids had determined that depending on the exact measures applied, a greatly varying proportion of the models passed the z-scores symmetry test. The results of the analysis of the actual pattern of internal zonation of the earth passed the symmetry test and produced mean r scores of as low as 0.001, a figure so close to zero as to be surpassed in this respect only a handful of times across the hundreds of thousands of random numbers simulations I have performed. In the latest analysis [9], mean r values of 0.003 were obtained, not only for the present configuration of continents and oceans, but also for two other periods in geological history extending backward some 200 million years. These were compared to an idealized earth structure model in which the 'surface zones' were averaged out to create a zone of consistently equal depth, in contrast with actual conditions (in which continental crusts, especially, cause a zone of greatly varying depths around the world). The idealized structure also passed the symmetry test, but the mean r values, though still very small, were about twice those connected with the actual earth patterns. These results are quite suggestive, but will require a sampling grid of points within the earth much finer than I was able to apply (involving between 15,000 and 20,000 internal earth locations) to fully settle the matter.

4 Discussion

My efforts at developing this model, continued over a more than 30-year period, have been slowed by a number of factors, not the least of which is its generally 'foreign' feel. Yet, its investigation does not seem to violate any basic tenets of the methodology of science, and so far, the results obtained have not been sufficient to make me think that efforts in this direction should be shut down. Some of these results, especially from the stream basins work, appear difficult to dismiss that easily.

Assuming for the moment that this Spinoza-inspired understanding of the way complex systems are organized is actually (at least, to an approximation) correct, a number of very important applications can be imagined. Philosophically, of course, this is a model of nature invoking final causes. But the final causes imagined are quite distinct from the old teleology, dependent on notions of supra-normal intervention toward predestined ends. Instead, systems are imagined to emerge as temporary spatial structures from their antecedents, only to exhaust their potential at some point and be re-absorbed by their surroundings (in the case of living things, 'die'). Their organization may be based on prior programs (as in the case of DNA-based life), but within that program there is the flexibility to change, so long as that change results in novel structure still observing the fundamental constraints on subsystemization suggested here.

As an evolutionary model, therefore, this understanding is distinct from Darwinian doctrine in steering away from anything describable in simple 'random walk' terms. Yes, an infinite number of potentialities of development remain in any given instance, but a 'much smaller' infinite number than we now assume to be the case.

As a practical matter, it must be remembered that to hold water, the model here must pertain to literally everything that can be construed as a naturally developing structure (including elements of human civilization, though not extending so far as providing an ultimate meaning for vacuum cleaners!). This means not only at the mesoscale we are most familiar with, but also with the very large and very small. It is perhaps not irrelevant, for example, that the fundamental interactive forces of physics and the codon genetics of the DNA molecule are each connected to elemental phenomena that are four in number.

For the present, it seems incumbent to develop one or two applications that will really serve to focus attention. Stream basins, though perhaps an ideal starting point from the point of view of measurement, will ultimately not accomplish this; a more detailed look at the spatial configuration of the earth's internal zones might serve better. But best of all would be some medical application, possibly as related to medical imaging. The human brain, for example, exhibits a lot of properties that are not dissimilar to the potential energies present across a drainage basin, and might be measured in analogous fashion. Brain temperature, electrical activity, blood flow rates, etc. are general properties all of whose measures could be sampled spatially in just the fashion I have described here, and subjected to analyses designed to expose their patterns of operation. It would be supposed that a fully healthy brain, analyzed to the ends I have discussed, should produce a final mean r measure little deviating from zero. Hypothetically, the more diseased and/or degenerating the structure, the more it should measure out to produce a mean r statistic of higher value. Concerted study in this direction might make it possible to identify, for example, susceptibility to, or early onset of, Alzheimer's disease. A research strategy in this direction is easy to imagine.

Possible applications of the 'maximum likelihood tree' model are less easy to imagine, but even there such an approach might prove useful to understanding, say, the pattern of ontological development of organ systems in development, or even to the pattern of ongoing emergences of ocean crust plates in plate tectonics.

I do not claim that the materials I have presented here and elsewhere prove the validity of this model. But, in fact, even if subsequent studies do not bear it out, the more basic idea of a 'final causes' approach toward complex systems it suggests should also provoke thought, as it introduces a way of understanding how system-level constraints on 'random walk' processes may operate.

References

[1] Wilson, A.G., Ecological and urban systems models: some explorations of similarities in the context of complexity theory. *Environment and Planning A*, **38**, pp. 633–646. 2006.

[2] Wilson, A.G., Boltzmann, Lotka and Volterra and spatial structural evolution: an integrated methodology for some dynamical systems. *Journal of the Royal Society Interface*, **5**, pp. 865–871.
[3] Smith, C.H., A system of world mammal faunal regions. II. The distance decay effect upon inter-regional affinities. *Journal of Biogeography*, **10(5)**, pp. 467–482.
[4] Carnap, R., *Der Logische Aufbau der Welt*. Weltkreis-Verlag: Berlin-Schlachtensee, 1928.
[5] Quine, W.V., Two dogmas of empiricism. *Philosophical Review*, **60**, pp. 20–43, 1951.
[6] Smith, C.H., A general approach to the study of spatial systems. I. The relational representation of measurable attributes. *International Journal of General Systems*, **12(4)**, pp. 359–384, 1986.
[7] Wilson, A.G., *Entropy in Urban Planning*. Pion: London.
[8] Smith, C.H. & Derr, M., 'In space' or 'as space'?: a new model. *Life*, **2(3)**, pp. 243–254, 2012.
[9] Smith, C.H., 'In space' or 'as space'?: spatial autocorrelation properties of the earth's interior. *International Journal of Geosciences*, **5(4)**, pp. 375–382, 2014.
[10] The Once and Future Wallace: A Theory of Spatial Systems, Online. http://people.wku.edu/charles.smith/once/writings.htm#2

Toward thermodynamics of real-time scheduling

I. Mayorov[1] & P. Skobelev[2,3]
[1]*Samara State Technical University, Russia*
[2]*Smart Solutions, Ltd, Russia*
[3]*Samara State Aerospace University, Russia*

Abstract

The modern problem of real-time resource management to increase enterprise efficiency is considered.

A new look at the dynamic self-organizing processes based on multi-agent technologies in building and revising schedules by events in real time is suggested. Schedule is considered as a flexible network of operations of demand and resource agents. This schedule is formed during the interactions of basic agent classes that set and break the dynamic links between each other, depending on the events and changing situation in the real world.

A thermodynamic model of demand–resource network (DRN) dynamics is introduced. There is a similarity to Ilya Prigogine's non-linear thermodynamics theory which allows us to explain the phenomenon of unstable equilibrium emergence, order and chaos, catastrophes, bifurcations and other non-linear events that are significant to the self-organizing processes control in multi-agent systems (MASs).
Keywords: adaptability, chaos and order, complex systems, demand–resource network, multi-agent technology, network dynamics model, non-equilibrium, real-time scheduling, self-organizing.

1 Introduction

Global economy's complexity is well known nowadays, which is often driven by uncertainty and dynamics in demand and supply.

Under these circumstances, the more operative, flexible and efficient approaches to the decision-making for resources allocation, scheduling,

WIT Transactions on State of the Art in Science and Engineering, Vol 90, © 2016 WIT Press
www.witpress.com, ISSN 1755-8336 (on-line)
doi:10.2495/978-1-78466-155-7/007

optimization, coordination and control are needed. These approaches allow the enterprises to maintain their profitability by avoiding equipment downtime or shortages of all kinds of resources, including personnel, equipment, finance and others. Therefore, the idea of a 'real-time smart enterprise' is discussed more and more often. These enterprises will have 'on-the-fly' dynamic resources scheduling in real time, since decisions' quality and efficiency directly depend on the moment of time itself. Moreover, in the conditions of today's competitive environment, delayed decisions lead to the losses.

A new approach to scheduling is suggested in this paper. In this approach, resources schedule is not a static data structure that is created only once and becomes outdated immediately, but it is considered as a dynamic self-organized data structure that adaptively changes. Unlike our earlier papers dedicated to the DRN creation for adaptive intelligent resource scheduling system based on the multi-agent technologies [1, 2], in this paper we suggest and justify a new 'thermodynamic' interpretation of self-organization processes of scheduling that shows a deep analogy of these processes to the non-linear thermodynamics suggested by Prigogine and Stengers [3] and Prigogine and Nicolis [4].

There is a gap between the existing solutions built based on the combinatorial method and scheduling tools [5], where we need to know all the orders and resources beforehand. Traditional systems implement batch versions of linear or dynamic programming, constraint programming and other methods based on combinatorial search of options [6, 7], which in practice turn out to be unsuitable. In contrast with the classical large, centralized, monolithic and sequential programs, multi-agent systems (MASs) are built as distributed communities of small autonomous software objects working asynchronously but in a coordinated way to obtain the results.

Multi-agent technology is considered as a new paradigm for solving complex problems that are difficult or even impossible to solve by classical mathematical methods or algorithms [8], for example, in scheduling and optimization, pattern recognition, text understanding and some other domains. This paradigm is considered as one of the most innovative and efficient for real-time scheduling of resources 'on the fly' [9, 10].

In this paper, we will consider the basic concepts of the approach and show its applications as well as a direct link with models of thermodynamics.

2 Development of the concept of DRN

The concept of DRN was suggested in our early papers on multi-agent approach. This concept was successfully applied in the creation of a gallery of adaptive real-time schedulers [1, 2, 7].

The DRN model is based on a 'holonic' approach to the production systems creation [11] where specific classes of agents are introduced.

As a part of the holonic approach in the DRN, we introduce *demand and supply agents*. These agents can act as sub-agents (assistants) to the main agents listed above, allowing them to concurrently or asynchronously search for each other on the virtual market, performing a continuous matching (Table 1).

Table 1: Main classes of agents.

Agent class	Specification of agent behavior, main goals and tasks	Attributes
Order agent	The goal of the order agent is to complete the order in time with maximum quality, minimal cost, best delivery time and minimum risk. Creates operation agents corresponding to the technological operations or business processes	Defines service level, real and virtual money balance for order execution, given the specifications for resources, operations interdependencies, deadline for order execution, risks
Product (result) agent	The goal is to get the best result to match order specifications and requirements	Domain-specific product requirements which are specified in order
Resource agent	The goal is to maximize resource workload, get the best orders, tightly schedule the executed operations or provide continuous operation and specified requirement execution	Availability at any moment of time, scale of charges of usage, working capacity and maintenance, efficiency
Operation agent	Coordinates performance of specific technological operations and business processes, their scheduling on resources, searches for the best resources for execution, negotiates about sub-tasks allocation, proactively improves the situation in the schedule	Preferable and real allocation time, virtual and real accounts, current penalties and bonuses, processes and operations deadline
Staff agent or enterprise agent	Controls enterprise key performance indicators and balance workload of the resources, discovers the 'bottlenecks' in the schedule and generates overcoming strategies, seeks to increase profit and other specified indicators	List of available resources, preferences and constraints for this task or organization
Demand agent	Allows to find a supply (resource) by the requirement, using requirements specification	List of situational matching parameters
Supply agent	Allows to find a demand by the requirement, using requirements specification	List of situational matching parameters

A constant matching between the competing and cooperating demand and supply agents on the system's virtual market allows to build a solution to any complex problem as a dynamic network of connections between tasks (operations) that is easily modified in real time.

In the DRN concept, any physical or abstract entity agent of problem domain can create agents of demand or resources. As a result, the schedule can be formed as a kind of requirement-driven network of operations, which can be easily adapted by events in real time [1, 2].

Continuous improvement of the system key performance indicators is based on the compensation method, where agents, whose condition became worse, receive compensation in the form of virtual money from agents that improved their schedule.

Specific DRN-based methods and tools were developed to design adaptive MASs for real-time scheduling [12].

2.1 Thermodynamic DRN scheduling model

Even a small organization schedule is a much more complex and dynamic object than it may seem at first glance because self-organizing processes that are similar to the ones described in Ilya Prigogine's writings began to play an increasing part in scheduling in our systems.

In general terms, it is suggested to consider the schedule of a complex system as 'unstable equilibriums', which differ in strength in different directions. A new order brings in money that plays the part of the systems energy. Order and chaos phenomena, autocatalytic reactions, oscillations and other non-linear events arise in the system. If the conflicts resolution and the number of messages in the system increase, we can assume that the temperature of the schedule section is increasing as well. Number of messages related to number of nodes plays the role of the system temperature. To support the structure of schedule, agents are paying some taxes that dissipate part of the input energy.

If orders stop coming in, this schedule will begin to gradually disintegrate in some time, from the weakest links between agents to the system transition to a state of complete chaos with zero energy, and cool down completely.

2.2 Satisfaction and formal problem statement description

The formalized problem statement is based on searching for a consensus between agents in DRN virtual market and can be formulated as follows.

Let us assume that all agents of demands and supply have their own goals, criteria, preferences and constraints (e.g. due date, cost, risk, priority, required equipment type or worker qualification). The importance of each criterion can be represented by weight coefficients in a linear combination of criteria for the given situation in scheduling, but can change during the schedule forming or - execution.

Let us introduce the satisfaction function for each agent (Fig. 1a), which will show deviation of the current value of this function from the given ideal value by any of the criteria for the current step of finding scheduling solution for this agent. The activity of agents also depends on bonus/penalty function and current budget allocated on specific accounts for virtual money (Fig. 1b).

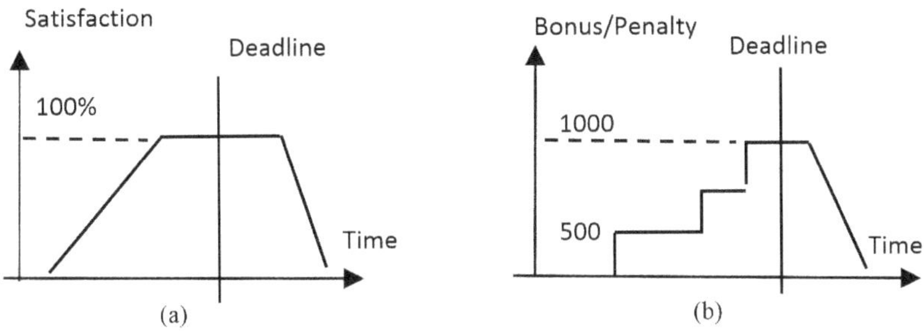

Figure 1: Example of (a) satisfaction function and (b) bonus/penalty function.

Let each demand j has several individual criteria x_i and suggested ideal values x_{ij}^{id}. For each agent of demand j, a normalized bonus/penalty function is calculated by component I, given, for example, as a linear function $f_{ij}^{task}(x_i - x_{ij}^{id})$. In most cases, this function has a bell form with maximum in the point of suggested ideal value. After that, a satisfaction value for each task agent and each criterion i with the given weight coefficients a_{ij}^{task} task is estimated.

By the proper selection of signs and form of the function, the goal of agent system can be reformulated as maximizing of virtual value y_j^{task} of demand j (upper index task means that the values belong to the demand agents):

$$y_j^{task} = \sum_i a_{ij}^{task} \cdot f_{ij}^{task}(x_i - x_{ij}^{id}) \tag{1}$$

where $\forall j$ weight coefficients are normalized: $\Sigma_i\, a_{ij}^{task} = 1$.

Similarly, the problem of finding the states x_{ij}^{*} of agents of demands j that maximize the total value of all orders can be formulated:

$$\begin{aligned} y^{task} &= \Sigma_j\, \beta_j^{task} y_j^{task} = \Sigma_j\, \beta_j^{task}\, \Sigma_i\, a_{ij}^{task}\, f_{ij}^{task}(x_i - x_{ij}^{id}) \\ y^{task*} &= \max_{xi}(y^{task}) \end{aligned} \tag{2}$$

where β_j^{task} is demand weight that allows to set and dynamically change the priorities showing the importance of criteria.

Similarly, the value function can be given for the supply by criteria z_k, with bonus/penalty function $f_{kl}^{res}(z_k - z_{kl}^{id})$, weight a_{ki}^{res} of criterion k for resource l, and resource value β_i^{res} for the system (which is similar for weight for demand agents function):

$$\begin{aligned} y^{res} &= \Sigma_i\, \beta_i^{res} \cdot y_i^{res} = \Sigma_i\, \beta_i^{res}\, \Sigma_k\, a_{ki}^{res} \cdot f_{ki}^{res}(z_k - x_{ki}^{id}) \\ y^{res*} &= \max_{zk}(y^{res}), \end{aligned} \tag{3}$$

$$z_k \in D^k, x_i \in D^I \ \ \forall i,k,I = Dim(D^I), K = Dim(D^k). \tag{4}$$

Variables x and z belong to some areas of the space of criteria for demands and supply, I and K are dimensions of the corresponding criteria spaces, and upper index res means that the values belong to resource agents.

Thus, in DRN, the optimization problem is formulated as solving eqns (2)–(4).

In other words, in the suggested bottom-up methodology, one global optimizer is replaced by many small local optimizers which are able to negotiate and find trade-offs when they search their local optimums.

The process of computations will be stopped and solution of problem will be delivered when consensus between agents is found.

2.3 DRN dynamics

Fundamental principles that form the basis of a new thermodynamic model of a DRN are the following:

- Each of our systems is open, which means that new orders and resources arrive at any moment of time, causing new agents' appearance and changes in the internal configuration of the connections between them in the system.
- Each new order brings real money into the system that plays the part of 'energy' required for the support of agents' activity (dynamics) and reconnection (statics), and part of this energy dissipates (stored in the system).
- Each agent is given a satisfactory function as its objective function, which shows how close the agent is to the specified ideal values by a certain criterion or their convolution, to which the agent converges.
- The further the agent is from the ideal value, the greater is the agent's 'aspiration' to achieve the results, the more active is an unsatisfied agent and the less active is a satisfied agent. This allows to control agents' internal activity in the system, which is particularly required for complex schedules design with hundreds of thousands agents.
- Agents are awarded if their state improves toward the set goal and get penalties otherwise. For that purpose, a special microeconomics of considered systems, where agents receive virtual currency and pay for a position in the schedule, changes in schedule or communication with other agents and taxes for the specified activity types. This is a dissipative part of the system.
- Each agent has a current virtual account and financial resources that are used by agents to improve their local allocation in the schedule.
- Agents iteratively improve their criteria in order to reach the best values that are close to ideal, compensating other agents' losses in concessions from their budget.

DRN model can be a powerful tool for self-organization processes' description and research in MAS.

It is convenient to introduce the following concepts for new system phenomena description connectivity, connection strength and autonomy level [13].

The connectivity of agents denotes the degree to which an agent is connected to other agents in the system. *The higher the connectivity of agents, the greater is complexity of the system.*

The connectivity of agents can flexibly change during the computations and at first, it limits agent choices by the closest environment, but the local interaction area is gradually growing unless there is no better solution.

In fact, it allows to control the speed and quality of a problem solution, increasing or reducing opportunities for self-organization in the system. *Connection strength* between agents denotes the mutual agent's satisfaction of the connection and degree of breakability of connections.

Weaker connections are easier to break and make new connections. Obviously, connection strength between agents can be calculated as an average satisfaction with connection. That causes misbalance of connections and that increases complexity and, therefore, the unpredictability of global behavior. *The weaker the inter-agent connection, the greater is complexity of the system.* Similar to the thermodynamics theory, the easier connections and, therefore, the more flexible the system is, the closer it is to the chaos and vice versa, with stronger connection between the agents system reaching a strict order state. Autonomy of agents denotes the degree of freedom given to them to decide what to do. *The higher the autonomy of constituent agents, the greater is complexity of the system.*

DRN model, topology and connection strength allow us to introduce order and chaos characteristics in MAS.

Lesser the connections, more is the chaos, and the more the connections and their strength, the higher is order degree. A number of connections of one agent can be used for order degree evaluation. Quantitative characteristics and their dynamics are the subject to our future investigations.

Demand agents establish links with resource agents using virtual money as payment for connection. If it is necessary, other agents' deteriorations can also be compensated with virtual money. Therefore, virtual money in such processes represent the energy that agents can use to improve their states. After the rescheduling, agents have some money left and the proportion of free energy decreases. Over time, agents' possibilities to change the schedule with compensations decrease, connection costs increase and their relevance in the sense of a positive impact on the degree of satisfaction of the system is lost. Staff agent can determine which of the connections only inflict losses and that it is energetically profitable to destroy them and plan the costly schedule parts. Therefore, some schedules 'crush' at some point.

New order arrival brings in free energy for the schedule changes. Schedules deterioration can be caused by the energy (virtual money) dissipation in the system. That corresponds with the open systems behavior in Prigogine's thermodynamics.

The connectivity, connection strength and autonomy of agents provide an opportunity to consider the evolution of interacting agents in the DRN using thermodynamic description. This description is similar to the autocatalytic reactions' description in complex multicomponent systems.

3 Self-organization in complex systems

The general behavior of complex systems is composed of agents' interactions, which in turn constrain their behavior. This behavior is called emergent, and it is unpredictable, but not chaotic. Emergency leads to the events associated with the dynamic rescheduling, as well as to the system adaptability to the impact of external influences.

3.1 Measuring adaptability of MASs

MAS receives orders that are scheduled for execution on resources. Incoming and previously distributed orders are dynamically reallocated to the system resources, since the agents tend to increase their satisfaction.

When a new order arrives and it is not distributed by the system yet, system satisfaction drops as the new agent cannot find the best allocation immediately. Over time, the overall satisfaction increases by rescheduling and gradual improvement of agents' conditions. The system goes into a non-equilibrium state, and then the agents try to find new local balance. It is suggested to calculate an average satisfaction for y tasks and resources agents depending on the time, in order to evaluate the dynamics of MAS:

$$y(t) = \frac{\sum_j y_j^{task}(t) + \sum_l y_l^{res}(t)}{M(t)+N(t)}, \tag{5}$$

where y_j^{task} is the task j agent satisfaction, y_i^{res} the resource l agent satisfaction, and $N(t)$ and $M(t)$ are the number of task and resource agents correspondingly. The tasks come into the system and the resources can be turned on and off; therefore, their number depends on time.

Let us introduce a degree of adaptability γ of an MAS [14]. This degree characterizes the rate of local equilibrium recovery:

$$\gamma = (y_2 - y_1)\cdot\frac{1}{T}, \tag{6}$$

where y_1 is a minimum satisfaction value after the influence, y_2 is average system agents' satisfaction after the MAS was influenced and T is the time of equilibrium recovery of the average satisfaction y_2.

A similar effect of a partial recovery can be observed not only when the resources are turned off but also with the abrupt appearance of new task flows. The higher the degree of adaptability, the greater the ability to agents' self-organization in elimination of disruptive influences is.

It is clear that MAS adaptability depends on the external influences intensity; for example, with the arrival of large task flows that exceed system productivity, the satisfaction decline will only grow with time, as the dissatisfaction increase of the arrived agents will not be covered by the partial growth due to rescheduling. Dynamics of the MAS satisfaction recovery during resources disabling in a time equal to 18 points are shown in Fig. 2.

After the average satisfaction maximum decrease to the y_1 level at time T, MAS finds a new quasi-equilibrium state y_2. $\Delta z = y_{max} - y_2$ is an irretrievably lost

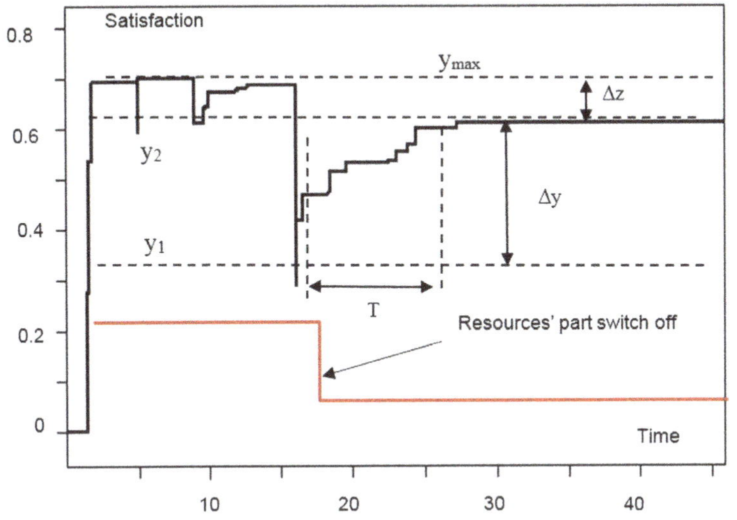

Figure 2: Adaptive partial recovery of the characteristics of a real MAS.

satisfaction. Similarly, we can consider task agents adaptability γ^{task} and resources agents adaptability γ^{res}.

Therefore, DRN degree of adaptability can reflect the degree of 'adaptive intelligence' as a measure of agents' self-organization.

3.2 State of unstable equilibrium and disaster

During their evolution, MASs discover specific events such us dynamic formation of non-equilibrium structures of different depths of inclusion and configuration (like formation of order from chaos) represented by spontaneous acceleration of agent negotiations, leading to a 'constructive destruction' of a schedule in order to create a new schedule of higher quality. Spontaneous acceleration can occur without any apparent cause, akin to autocatalytic processes [4]. The acceleration usually leads to the accumulation of virtual money resulting in a kind of explosion or catastrophe (radical changes in the schedule). Let us consider this on the example of transport logistics problem in Fig. 3.

The scene depicts a community of several trucks, each in turn containing a community of journeys, each of which containing a community of cargoes. Each link between demands and resources is labeled with two figures denoting the perceived values of the link by both nodes connected by the link.

We can see that the order 1 is satisfied with the allocation of resources. Order 2 provides for less satisfied and not satisfied agents a new opportunity to improve their allocation. Many negotiation processes will begin which are proactively initiated by resources that aim to improve their allocations.

This increased activity of agents, sensing new opportunities combined with their dissatisfaction with the decision by the truck agent to accept new cargoes

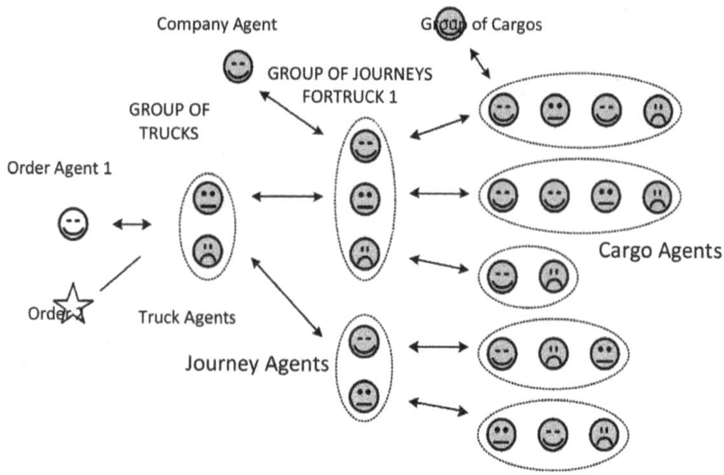

Figure 3: An example of autocatalytic reactions in the scheduler.

and create new journeys, may result in a ripple effect of changes to the schedule, accelerating the rate of change and causing a full collapse of the previously agreed schedule and its immediate re-building in a new manner. The schedule thus passes through a slowdown, accelerated activity, collapse into chaos and re-birth, a process known as 'constructive destruction'.

3.3 Improving deteriorations

In cargo transportation simulation system [15], orders are executed with a possible delay (penalties); transportation cost of each truck per time unit is the same. Unavailability period is simulated for one of the trucks (e.g. the truck is not taking orders). The simulation was performed to determine the total profit of the system, including penalties and empty runs of trucks to the orders loading points (Fig. 4).

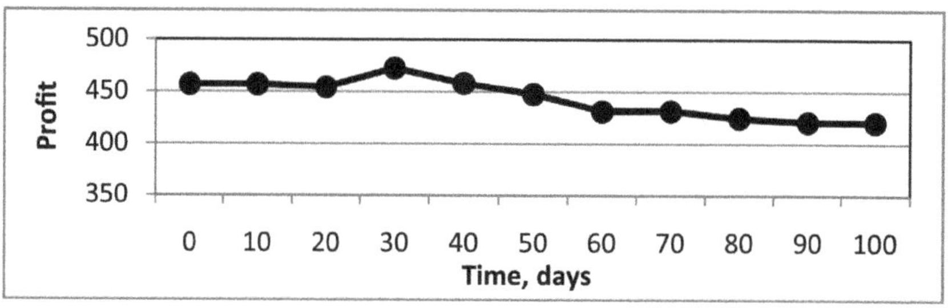

Figure 4: Transportation system total profit depending on the unavailability time of one of the trucks.

The average profit decreased; however, in the 30-day period of unavailability, a 5% growth was observed, that is, 'situation deterioration' in this case led to more successful non-equilibrium state of the schedule, which is associated with a more profitable orders allocation to the remaining trucks. This phenomenon is similar to the well-known Braess's paradox [16]. In our case, adding trucks' working hours worsens the results and the introduction of restrictions improves it. In this case, adding a new resource worsens the results and the introduction of restrictions improves it. A multi-agent cargo transportation system increased its profit at some time period with decreasing part of the resources by self-organization and rescheduling.

3.4 Loss of causality

The initial events that lead to the rescheduling activate order and resource agents. Branching networks of changes do not allow identifying an event or an agent that caused the changes. Therefore, in time, the agents forget in what stages if the rescheduling they were involved. Waves of disturbance in the system activate the branching processes that do not contain the initial information about the cause of changes.

There were from two to six rescheduling times during the simulation of the loading distribution on the executing elements even with a small number of tasks (Fig. 5).

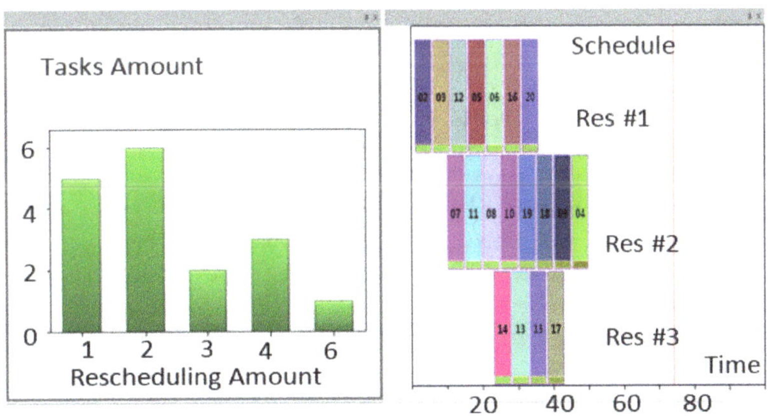

Figure 5: A bar chart of the system rescheduling (3 resources, 20 incoming tasks).

4 Conclusions

Using the proposed DRN model, it will be possible to examine in detail in the future research the dynamics of self-organization processes, depending on the external incoming events, changes in the resources availability and operation in the condition of the changing microeconomics of agents system.

Metrics for the complexity parameters evaluation of the DRN will allow to study the dependency of adaptive properties of agents' self-organization from the dynamic structure of the network and to predict its response to external influences.

Dynamics model will provide an opportunity to study the processes of development and deterioration of the schedules over time as a phenomenon similar to the non-equilibrium thermodynamics.

Acknowledgement

The study was supported by the Ministry of Education and Science of the Russian Federation.

References

[1] Skobelev, P., Vittikh, V. Models of Self-organization for Designing Demand-Resource Networks, Automation and Remote Control. *Journal of Russian Academy of Science*, **(1)**, pp. 177–185, 2003.

[2] Vittikh, V., Skobelev, P. The compensation method of agents interactions for real time resource allocation, Avtometriya. *Journal of Siberian Branch of Russian Academy of Science,* **(2)**, pp. 78–87, 2009.

[3] Prigogine, I., Stengers, I. Order out of Chaos: Man's new dialogue with nature. Flamingo, 1984.

[4] Prigogine, I., Nicolis, G. *Self-Organization in Non-Equilibrium Systems.* Wiley. 1977.

[5] Leung, Y.-T. *Handbook of Scheduling: Algorithms, Models and Performance Analysis, CRC Computer and Information Science Series*, Chapman & Hall, London, 2004.

[6] Shirzadeh Chaleshtari, A., Shadrokh, S. A Branch and Bound Algorithm for Resource Constrained Project Scheduling Problem subject to Cumulative Resources, *World Academy of Science, Engineering and Technology*, **6**, pp. 23–28, 2012.

[7] Wooldridge, M. *An Introduction to Multi-Agent Systems*, John Wiley & Sons, London, 2002.

[8] Shoham, Y., Leyton-Brown, K. *Multiagent Systems: Algorithmic, Game-Theoretic, and Logical Foundations*. Cambridge University Press, New York, 2009.

[9] Pechoucek, M., Marík., V. Industrial deployment of multi-agent technologies: review and selected case studies. *Autonomous Agents and Multi-Agent Systems*, **17(3)**, pp. 397–431, 2008.

[10] Leitao, P., Vrba, P. Recent Developments and Future Trends of Industrial Agents, *Proceedings of the 5th International Conference on Holonic and Multi-Agent Systems in Manufacturing*, Springer, Berlin, pp. 15–28, 2011.

[11] Brussel, H.V., Wyns, J., Valckenaers, P., Bongaerts, L. Reference architecture for holonic manufacturing systems: PROSA. *Computer in Industry*, **37(3)**, pp. 255–274, 1998.

[12] Skobelev, P. Multi-Agent Systems for Real Time Resource Allocation, Scheduling, Optimization and Controlling: Industrial Applications (Invited Talk). *Proceedings of the 5th International Conference on Industrial Applications of Holonic and Multi-Agent Systems*, **6867**, pp. 1–14, 2011.

[13] Rzevski, G., Skobelev, P. *Managing complexity*. WIT Press, UK-USA, 2014.

[14] Zadeh, L.A. On the definition of adaptivity. *Proceedings of the IEEE*, **51(3)**, pp. 469–470, 1963.

[15] Granichin, O., Skobelev, P., Lada, A., Tsarev, A. Cargo transportation models analysis using multi-agent adaptive real-time truck scheduling system. *Proceedings of the 5th International Conference on Agents and Artificial Intelligence (ICAART'2013)*, February 15-18, 2013, Barcelona, Spain. – SciTePress, Portugal, **2**, pp. 244–249, 2013.

[16] Braess's paradox. [online] Available at: https://en.wikipedia.org/wiki/Braess's_paradox [20 January 2015].

Semantic shift to pragmatic meaning in shared decision making: situation theory perspective

M. Canan, A. Sousa-Poza & S. F. Kovacic
Engineering Management & System Engineering, Old Dominion University, USA

Abstract

The way humans establish communication depends on the generation and conveyance of meaning. Linguistically, meaning in information is dependent on the meaning that is ascribed to signifiers in the context of the communication. These signifiers can include items such as words, phrases, signs, and symbols. Conveyance of meaning may, however, imprecise and prone to error.

The meaning of information in communication may arise from a change in the context in which a signifier is placed (intrinsic), or a change in the paradigm with which the signifier and context are perceived (extrinsic). In simple situations, where paradigms are reconcilable, semantic shift is solely intrinsic. In complex situations, where differing paradigms will generally lead to irreconcilable perspectives (paradoxes and dualities); the semantic shift will be both intrinsic and extrinsic. Decisions are based on an individual's (or individuals' shared) understanding and understanding is in turn contingent on perspective. Decision making will, therefore, be affected by discrepancies in meaning.

It is critical to understand the nature of the discrepancies where shared awareness is necessary to enable group decisions. The theoretical construct presented recognizes that (1) a semantic shift may be required where multiple perspectives based on different paradigms come into play and (2) a semantic shift may introduce error, inefficiency, noise or redundancy. Therefore, individual limits can be recognized via shared awareness, which can be studied with situation theory.
Keywords: semantic, pragmatics, situation theory, situational awareness, shared awareness, decision making.

WIT Transactions on State of the Art in Science and Engineering, Vol 90, © 2016 WIT Press
www.witpress.com, ISSN 1755-8336 (on-line)
doi:10.2495/978-1-78466-155-7/008

1 Introduction

The universal definition of decision making, namely: "choosing among alternatives", renders decision making an essential component of daily life. The spectrum of daily life is comprehensive, constituted by a variety of systems, such as military, private business, governments, or families. The two salient common characteristics of these systems are that (i) the primary agents in these systems are human beings and (ii) all these systems constitute a communication system that allows an information flow among the communicative components, or agents. Communication depends on the generation and conveyance of meaning. Linguistically, meaning in information is dependent on the meaning that is ascribed to signifiers in the context of the communication. These signifiers can include items such as words, phrases, signs, and symbols.

The meaning of information in communication may arise from differences in the context in which a signifier is placed (intrinsic), or a change in the paradigm with which the signifier and context are perceived (extrinsic). Regardless of being intrinsic or extrinsic in terms of cause, the differences entail discrepancy in meaning, which entails an epistemological gap. Therefore, the best possible decision is contingent on this epistemic gap, which separates the objective individual claims from objective communal claims.

This work is not a linguistic study. Rather, the primary concern is how an efficient informative communication can be attained. A Situations Theory perspective is utilized where situation is defined as a set of conditions that human beings expand on with the requirement that an individual is or becomes cognizant of the set of conditions [1]. Situation in this context bifurcates in two states: simple and complex, where the demarcation between the two is constrained by comprehensibility and understanding of the described situation.

The social facts in human life are being discussed and studied with various disciplines with an inquiry of how human beings create social structures that are cooperative and productive. In this regard, information flow is important for human beings who commune in social structures with assertoric commitments that may need further justification for decision making. Therefore, social interaction fosters individual's need to form organizations that constitute the following salient dynamics: communication, understanding awareness, and decision making.

2 Communication

Communication is not a mere act of imparting information via language. It constitutes components of assertoric commitments that are verified with social interaction. Real things and the assertoric individual claims about real things are opaque. Thus, further probing is necessary for individual and for communal understanding to develop. Human beings utilized language as a communication system. This system constitutes messages to be communicated, communicative agents, and a set of signifiers. An abstract example of communication in an organizational setting is that there is a sender who would like to convey an understanding to a receiver so that a shared awareness can be established between a sender and a receiver.

2.1 Communication models

Communication can be undertaken in many different ways. These include textbooks, public speeches, social media, telephone conversation, media, email, radio, TV, and so on. The models presented here constitutes three levels of communication: *intrapersonal communication* [2], *public communication* [3], and *interpersonal communication.*

The first two communication types are not representing the communal aspects of communication. However, these two form the basis for the interpersonal communication.

Communication models are a representation of certain aspects of events, structures, or system made by using symbols [4]. The salient communication models such as Lasswell [5], Shannon and Weaver [6], Berlo [7], Dance [8], Barnlund [8] conform a sender and a receiver interaction with different approaches. They all agree that, communication is dynamic, continuous, circular, unrepeatable, irreversible, and complex.

2.2 Systemic model of communication

Systemic approach to communication is utilized in modeling communication as a system. The characterization of communication with the General System Theory's primary principle is comprehension of the related parts so that a change in one aspect entails changes in all other parts [9]. The systemic approach to communication constitutes the following axioms introduced by Watzlawick and Jackson [10]; (i) the impossibility of not communicating; (ii) content and relationship in communication; (iii) the punctuation of the sequence of events; and (iv) symmetrical and complementary interaction.

2.3 Language as a communication system

Theorists often view language as a communication system. According to Palmer, language relates something to be communicated with something that communicates, a message on the one hand with a set of signs or symbols on the other [11]. Palmer maintains a reservation in classifying language as a communication system. Language is not composed of a message in the form of information, because of the fact that its function's framework is contingent on social relationships. Furthermore, the signified and signifiers' [12] relations in human context is complex, in contrast to the communication system of a traffic light. In these types of communication systems, message can be defined without a language constraint. For example, "RED" means stop.

2.4 Language as a complex adaptive system

In the human context, language manifests its characteristic as a social function. This manifestation has two manifolds; cognitive and physical functions which comprise a complex adaptive system (CAS). The system constituents are interacting agents. These agents render the system adaptive because of the fact that agents' behavior is based on past interactions, and current and past interactions together form future possible behavior [13]. The CAS framework renders language a dynamic, generative system rather than a static system of grammatical principles. According to Elis, this system has the following characteristics:

1. The system consists of multiple agents.
2. The system is adaptive; past and present behavior involves forming the future actions.
3. Competing factors affect the agents' behavior.
4. The structure of a language merges from interrelated patterns of experience.

Communication or language in human context can emerge at different points of social interaction. For example, joint actions and cooperative activities all require sharing human beings' intention, culture, with one another [14–16]. Therefore, the CAS approach permits studying language in human context in a way that the aggregated effect of many interacting constraints, including the structure of thought process, perceptual motor biases, cognitive limitations, and social-pragmatic factors [17], are taken into consideration.

2.5 Pragmatics in communication

Agents in communication aim to convey the information in a most efficient way. Besides, recognizing the tacit agreement, these agents utilize a variety of operative assumptions in the presupposition framework of the context of communication operation. The significance of these presuppositions is that they do not emerge from the context of the message, rather from contextually formed presuppositions, which is in the agents responsibility [2].

Extracting information is the essential operation in communication. This essentiality ensues from the fact that once the message is imparted, and the extraction (decoding) starts, the sender cannot intervene in this process.

2.5.1 Context in communication

The recipient controls the extraction of the information from the imparted message. The sender must in turn be careful in compiling the message to prevent the possibility of misunderstanding. The process of distinguishing, reducing, or even eliminating the pluralism in the constructed meaning of a verbally compiled message is called interpretation [2]. The process of interpretation utilizes context as a reference point. As Rescher says,

> in the process of interpretation context is not just important, it is everything.

Another salient characterization that is attained by context is about assumption and presumptions where both of them are contextual matters. The role of context is a principle of efficiency in communication.

2.5.2 Communication in communal setting

A viable and satisfactory communal existence requires that the people understand each other. From individual's point of view, understanding is an endeavor to render the preceding and the proceedings of a rational person, efficiently intelligible to the others in the communal existence. The communality of this rational procedure requires mechanisms such that cognitively confined knowledge can be efficiently probed via communication.

The probing phenomenology is the key element in progression of material science. In fact, human sciences utilize similar notion, however named differently. In human interaction framework, presumptions emerge in a context where humans have questions and seek answers. Yet, presumption is a thought instrument that is employed in circumstances where practical considerations are parameterizing and constraining human cognitive and communicative practices. For example, by changing presumptions, one can overcome a circumstance which lacking of evidence for a conclusive claim that precludes reaching an answer. Rescher's arguments on presumptions are as follows:

> Presumption affords useful cognitive and communicative resource. The obvious and evident advantage of presumption as an epistemic device is that it enables us vastly to extend the range of questions we are able to answer. It affords an instrument that enables us to extract a maximum of information from the communication situations. Presumption, in sum, is an ultimate pragmatic resource.

Presumptions are transient. They become substantial in the framework of verbal communication or written communication. Presumptions are contextual and issue driven; a putative fact that is widely employed in communication [2]. Thus, the notion of presumption gains interest in logics, semantics, and epistemology.

The problem stems from the fact that communication limits individual understanding, which is a contingency in making decision in complex environments. Individual limits can be recognized via awareness: a condition of having or showing realization, perception, or knowledge [18].

3 Situations theory perspective on understanding

Situations theory is a meta-theoretical construct that constitutes the cogitation of the circumvent condition of a problem. Therefore, rather developing higher understanding of specific phenomena from well-formulated perspectives, it focuses at understanding how such perspectives contribute to dealing with problems encountered in complex situation [1]. Consequently, here in this section we articulate understanding with in the scope of awareness, which is a situation theory perspective.

3.1 Situations theory and awareness

Situations theory facilitates the sustainment of natural ties to reality for any developed construct to understand dynamic, transient, and context-specific situations. Furthermore, situations theory enables the individual involvement of a participant construct rather than a mere observer [1]. This evolution emerged from:

- The incorporation of the participant and maintenance of the observer in the problems [19].
- The practical nature of the problem [1].
- The paradigm captured in Pragmatic Idealism that leads to the present state of understanding of situations theory [20].

The salient evolution for situations theory towards awareness entailed from the seminal work Pragmatic Idealism [20]. As an ontological construct, the formation of awareness is articulated in the Representation of Reality – RDP model.

3.1.1 Reality, domain, and perspective – RDP model

In defining a perspective P, a state of awareness is required. Establishment of a state of awareness is an extremely complex problem. The first issue is whether for awareness to exist we first need awareness of self, or whether we need awareness of other than self. To circumvent any debate of primacy, we view this as a mutually generative process (GP). The self defines that which we perceive and that which is perceived ultimately defines the self. Due to the imposition of the self on awareness, it must be recognized that every domain of awareness will in essence be unique, its formation defined by a person's dispositions, past experiences, and even future expectations. For instance, in an inquiry about another domain of awareness, the presuppositions ensued from these formations. The imposition of other than self on awareness will, however, also means that the perspectives, irrespective of individual interpretation, may share commonalities.

To be able to share perspectives in their totality; that is to say there is a sharing of meaning not only of information, two agents must share awareness. This produces an interesting dilemma. The assertoric nature of the formation process of each state of awareness would mean that sharing should not be possible. We, however, know this not to be true. To overcome this dilemma, the nature of awareness, and particularly the manner in which the domain of awareness is established must be studied further.

3.1.2 Awareness and the domain

The formation of a domain takes place through the GP. The GP inculcates the mutual generation of self-awareness and awareness of other than self (Fig. 1). The formalization of all that is other than self will at that point be captured as the domain of awareness. It is the basis of the situational construct; observer, entity, and solution form. Typically, complexity is a feature that results from the misalignment or tensions that exist between these three elements.

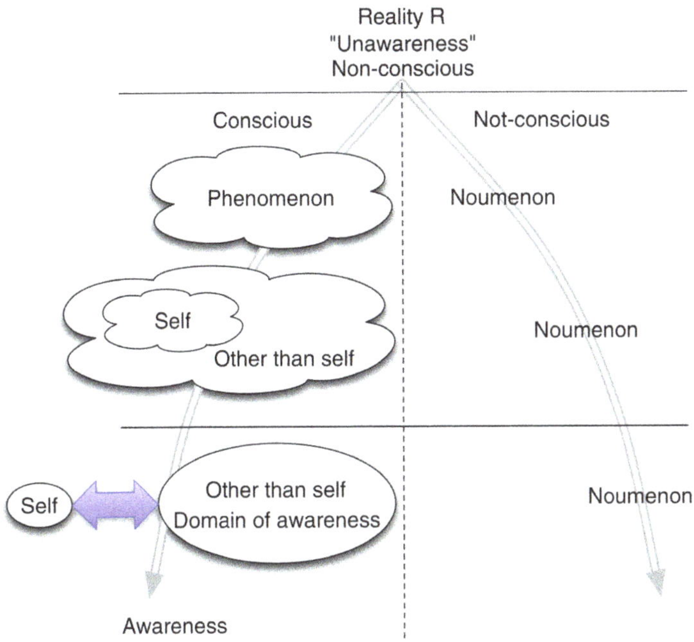

Figure 1: Construction of self and other than self. Noumenon represents unbounded participation of human agent, and phenomenon represents cognitive and bounded observation of life.

The domain of awareness plays a critical role in our ability to understand a problem. It captures the degree to which a problem may be comprehended. The degree to which abstractions in the form of simplifications and assumptions must be made will affect comprehension. The effectiveness of the domain of awareness can be represented by this abstraction in a nominal construct called the abstraction distance.

3.1.3 Comprehensibility, understanding, and complexity

For conditions where the domain is a crude approximation of the problem, as represented by the conceptual approximation distance $A'(d)$, the possible comprehension will be low. Conversely high comprehension requires a low approximation distance, where $A'(d) \rightarrow 0$. A perspective is formed within a domain of awareness. The perspective represents that which is understood of the problem. The perspective is itself an abstraction of the domain of awareness. Therefore, an additional abstraction distance ($A''(d)$) can be postulated that references the degree to which that which may be comprehended is understood. In this model, the complexity is a function of the effectiveness of what is understood. What is understood is the basis on which action is taken. Action is broadly defined to encompass any resultant activity. This may include a decision, a design, an explanation, or purposefully doing nothing. Notionally, action is derived from what are broadly referred to as *propositional statements*, or claims. Determinations of whether that which is understood is 'right' or 'wrong' can in a very rudimentary manner be defined by whether a claim is ultimately reflective of what transpires. The overall complexity of a situation that can be symbolized by A(d) is a combination of A′(d), and A″(d) (Fig. 2).

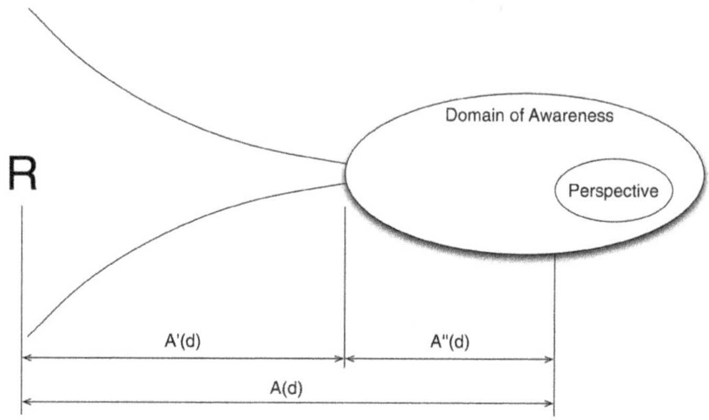

Figure 2: Domain of awareness and perspective with respect to reality.

In the complex situation, A(d) >> 0. The effect of this is that the perspective, which will be consistent with the domain of awareness, will only generate understanding of the problem that is extremely limited. This does not mean that a claim that is made will be wrong. The probability that the claim will be erroneous (CL(ε)) will [20], however, increase as A(d) increases. This does introduce a complication on the notions of right and wrong. Since in a complex situation, we should expect A(d) to always be high; the probability of making erroneous claims will always be high. In other words, the chances of being wrong are extremely high. A distinction in probability of making erroneous knowledge claims can be made based on whether the source is due to A′(d) or A″(d).

3.1.4 Generative process

The dynamics of GP that are incorporated into the situations theory renders the construction of self and other-than-self distinguishable from other domains and perspectives. The GP is how we come about forming a domain of awareness. At the highest level of resolution, every perspective will be completely unique based on the uniqueness of the components that form the GP (Figs 1–3).

The differences between two domains may lie not in the comprehension that they provide, but how that comprehension is generated. The generation process includes assertoric claims. These claims made by two domains may be the same, but not acceptable to each other based on how the claims were generated and how they are justified. The GP might be likened to a 'black box process' in which we know we are moving from a state of unbounded, non-conscious, complete connectedness *with* Reality to a state in which we have a bounded, cognitive, conscious perspective *of* Reality. The components in Fig. 3 can be summarized as; past experiences, coping strategies developed, and cognitive dispositions and cognitive predispositions.

3.1.5 Situational awareness

The SA definition provided by Endsley [21] renders situational awareness and integration of both process and product where the distinction between these two is that; (i) state of awareness representing the information and knowledge is product; (ii) perceptual and cognitive activities involved in constructing, updating, revising the state of awareness [22].

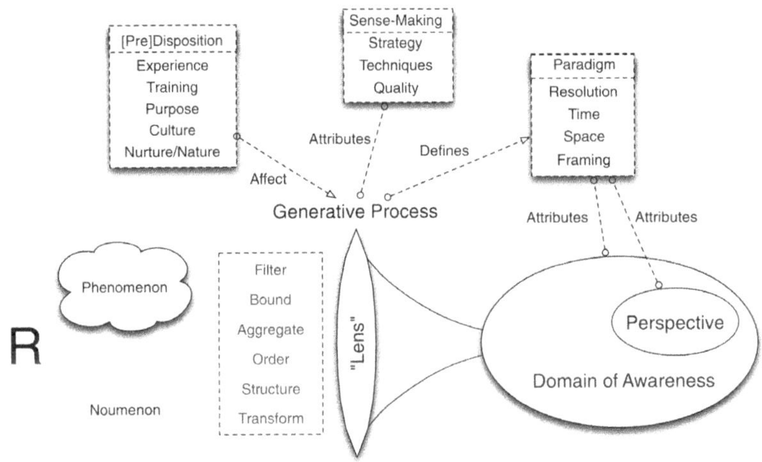

Figure 3: Components of generative process in the construction of a domain of awareness (is that which is comprehended) and perspective (is what is understood).

In this regard, the situational awareness is a construct of participant–observer duality that incorporates the integrated meaning stemming from spatial–temporal constraints, goals/purpose, understanding of the situation, perceived mental models, and classification of the situation. These are manifested in the interaction between the situation and datum in the construct of situational awareness. Since situation awareness, as a state of knowledge, is distinguishable from the GP [23], utilizing different GPs does not preclude maintaining common situational awareness. Furthermore, this GP constitutes a cogent model that allows anticipation and interpretation of the available information within the domain of awareness [24].

3.1.6 Shared situational awareness

The Situations Theory framework invokes the cognitive representation of reality (CRR) [25] for the representation of the cogent aspect of situational awareness. The significant contribution of the CRR concept is the shift from mere cognitive awareness to domain of awareness by utilizing the GP. Along with the RDP approximation, CRR (principle of awareness as the disassociation) approach mitigates constraints that emerge from awareness and knowledge relation. Therefore, any knowledge statement will have a degree of fallibility [18, 20, 25] that is inferential in a context that constitutes a new paradigmatic new worldview, generalizations of concepts, and placing the impact of participation into the situation. In this context, interaction of particulate domain of awareness becomes the medium for shared awareness [18] to emerge by utilizing purposeful information exchange.

3.1.7 Required condition for sharing

The a priori parameters required for the construction of congruent domains of awareness supporting conditional sharing include:

1. Link
2. Intentionality/imperative
3. Orthodoxy
4. Factors that affect the conveyance of meaning, or the establishment of congruent meaning
   a. Disposition
5. Transfer of factors influenced by disposition
   b. Signifiers and context
   c. Elements of the GP; how the context was formed, factors in the framing that will influence the manner in which interpretation took place
6. Metadata
   d. Time
   e. Place
   f. Nature of situation

In the simplest case where there is a link, intent, and no orthodoxy, this problem is reduced to a study of dispositions. How the claim was established by agent 1, and whether the claim is comprehensible, and can be understood by agent 2.

3.1.8 Type of shared awareness

Based on the General Theory of Shared Awareness [18], four general strategies for sharing are proposed depending on whether the shared space that is to be established will be derived from, or require conditional sharing, contextual sharing, synthetic sharing, or synoptic sharing. In this article, communication is studied only in the conditional sharing framework.

3.2 Shift in semantic to pragmatic meaning

Meaning is an emergent attribute of a message. Meaning is generated from the interaction of components of the message. The components will be tacit and explicit. The message will include content elements in the form of signifiers.

Meaning is derived from the manner that the signifiers are structured together and interact within a context. The context will include a variety of components including: The manner in which the message is conveyed, the manner in which the sender formulated the perspective from which the message is being generated, and even the manner that the sender expects the receiver to interpret or decode the message.

3.2.1 Truth meaning vs. used meaning

The meaning analysis of a statement in the context of linguistic is confined within the appropriateness and correctness of usage. Thus, the relation with real meaning is not commercial to articulate in communication models [26].

Communication generates integrated meaning that is contingent on context, time, predisposition, orthodoxy, etc. Since each of these contingencies renders the meaning idiosyncratic, RDP model can enhance communication model with distinguishing the truth meaning and used meaning where "truth" and "used" meaning defined as [26]:

u_{nm} (SA_i, P_{nm}) = the information shared situation SA satisfies the use condition for the claim P,

t_{nm} (S_i, P_{nm}) = the informative situation S satisfies the truth condition for the claim P.

Learning the meaning of a word (attaining an understanding/awareness of it) is a regimen of learning how it is used. Therefore, a shift in the meaning is possible that manifests itself in communication in the domain of awareness. This will take the form of discrepancies in the shared understanding that is established and will ultimately qualitatively influence shared/common decisions.

3.2.2 Communicating in conditional shared situational awareness

From a natural decision-making point of view, "choosing among alternatives" is elaborated as attaining a goal through reaching a common perspective. Shared awareness is a state of comprehension for a population that is functionally dependent on the establishment of shared awareness between agents. This form of sharing articulate one manner in which perspectives built on heterogeneous paradigms can form a conditional shared construct. In conditional sharing, one agent moves towards another agent's domain of awareness. Extrapolating the communication framework, the sender is viewed to be dominant. The sender is notionally viewed to have the desire, and predisposition to move the domain of awareness.

In Fig. 4, this is depicted in the RDP model. There are two domains of awareness, $D_1(t_n)$ and $D_2(t_n)$, multiple perspectives in these two domains of awareness P_{Dn}. The approximation distance to reality (R) for perspective is $A_D(d)$, for domain of awareness is $A'_D(d)$, and perspective with respect to domain of awareness is $A''_D(d)$. The separation between two domains of awareness is $A_s(d,t_n)$. There is a signifier domain for verbal communication where the constructed presumptions (= perspective) represent a meaning. This is indeed where the RDP model becomes important. According to the earlier definition of complex situation in [20], complexity is defined as "making an erroneous knowledge claim CL(ε)". The truth meaning – $t_{nm}(S_i,P_{nm})$ – is in the reality. When an agent initiates an approximation by making a claim of

$$CL_i(\varepsilon).\ t_{nm}(S_i,P_{nm}) \equiv u_{nm}(SA_i,\ P_{nm}).$$

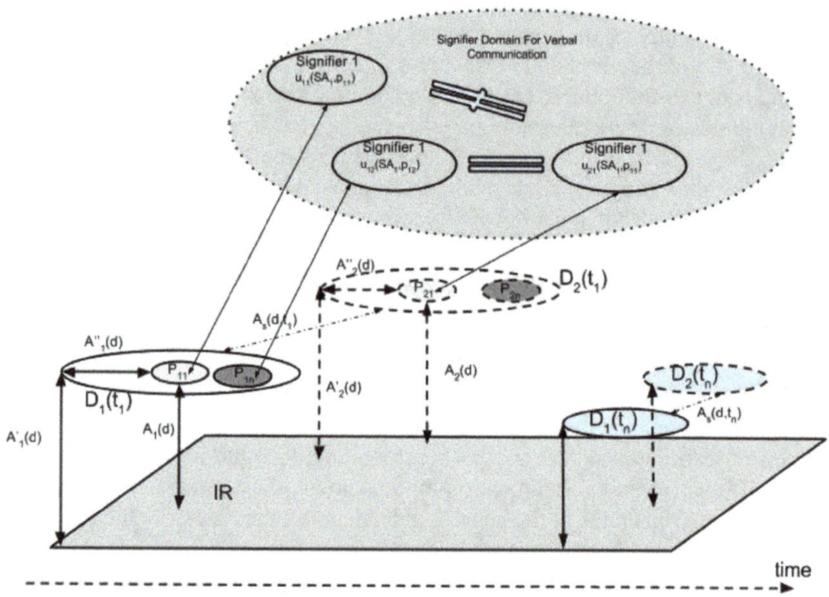

Figure 4: RDP representation of communication in shared awareness and communication.

Based on this construct, the integrated meaning is constrained by the intrinsic process. Explicitly, however, substantiation of the words is the same, which precludes attaining a shared decision.

The desire for sharing for the sender's domain, D_1, entails further $CL_i(\varepsilon)$ so that the same used meaning attained with move towards perspective in D_2

$$CL_i(\varepsilon).\ u_{nm}(SA_i,\ P_{nm}) \equiv u_{nm}(ShSA_i,\ P_{nm}).$$

This generates a shift in meaning from semantic to pragmatic to attain a decision. This algedonic loop that constitutes $CL_i(\varepsilon)$ iterations to the used meaning of the sender occurs so that the separation between the two domains of awareness $A_s(d,t_n)$ becomes $A_s(d,t_n) \rightarrow 0$.

4 Conclusion

The meaning of a word in linguistic framework is rooted in semantics, hence dominated over used meaning. However, the real meaning of a word can be understood possibly better with what come after it. Studying communication in situations theory perspective allows recognizing the proper causes of discrepancies in communal understanding. The pragmatic meaning, manageable, is tenable to overcome knowledge discrepancies that might causal emerge from different sources while constructing understanding. General Theory of Shared Awareness permits studying the communication in this framework so that predispositions and cultures involved can be preserved and adaptation required environments are sustained.

References

[1] Sousa-Poza, A., A narrative of [complex] situations and situations theory. *Managing and Engineering in Complex Situations*, Vol. 21, eds. S.F. Kovacic & A. Sousa-Poza, Springer: The Netherlands, pp. 13–44, 2013.

[2] Rescher, N., *Communicative Pragmatism*, Rowman & Littlefield Publishers, Inc.: Lanham, 1998.

[3] Huseman, R.C., Lahiff, J.M. & Hatfield, J.D., *Interpersonal Communication in Organizations*, Holbrook Press: Boston, MA, 1976.

[4] Chapanis, A., Men, machines and models, *American Psychologist*, **16**, p. 115, 1971.

[5] Lasswell, H.D., The structure and function of communication in society. *The Communication Ideas*, ed. L. Bryson, Institute for Religious Studies and Socia Studies: New York, 1948.

[6] Shannon, C.E. & Weaver, W., *The Mathematical Theory of Communication*. University of Illinois Press: Champaign, IL, 1949.

[7] Berlo, D.K., *The Process of Communication*, Holt, Rinehart & Winston: New York, 1960.

[8] Barnlund, D.C., A transactional model of communication, *Language Behavior: A Book of Readings*, eds. A.G. Johnnye Akin, G. Myers, J. Stewart, Mouton & Co. N. V.: Hague, 1970.

[9] Hall, A.D. & Fagan R.E., Definition of system. *General System* **1**, pp. 18–28, 1956.

[10] Paul Watzlawick, J.B.B. & Jackson, D.D., *Pragmatics of Human Communication: A Study of Interactional Patterns, Pathologies and Paradoxes*, Norton: New York, 2011.

[11] Palmer, F.R., *Semantic: A New Outline*, Cambridge University Press: London, 1976.

[12] De Saussere, F., *Course in General Linguistic*, McGraw-Hill: New York, 1959.

[13] Elis, N.C., Language is a complex adaptive system. *Language As a Complex Adaptive System*, Santa Fe Institute, 2007.

[14] Bratman, M., Shared cooperative activity. *The Philosophical Review*, **101**, pp. 327–341, 1992.

[15] Bratman, M., Shared intention. *Ethics*, **104**, pp. 97–113, 1993.

[16] Clark, H.H., *Using Language*, Cambridge University Press: Cambridge, 1996.

[17] Christiansen, M.H. & Chater, N., Language as shaped by the brain. *Behavioral & Brain Sciences*, **31**, pp. 489–558, 2008.

[18] Kovacic, S., Micro to macrodynamics of shared awareness emergence in situations theory: towards a general theory of shared awareness. *Engineering Management and System Engineering*, Old Dominion University, 2013.

[19] Sousa-Poza, A., Kovacic, S, & Keating, C., System of systems engineering: an emerging multidiscipline. *International Journal of System of Systems Engineering*, **1,** pp. 171–188, 2008.

[20] Sousa-Poza, A. & Correa, Y., Pragmatic idealism as the basis for understanding complex domains: the trinity and SOSE, *IEEE SMC,* Hawaii, 2005.

[21] Endsley, M.R., Design and evaluation for situation awareness enhancement. *Human Factors Society 32nd Annual Meeting,* Santa Monica, CA, 1988.

[22] Adams, M.J., Tenney, Y.J., & Pew, R.W., Situation awareness and the cognitive management of complex systems. *Human Factor*, **37(1)**, pp. 85–104, 1995.

[23] Endsley, M.R., Toward a theory of situation awareness in dynamic system. *Human Factor*, **37(1)**, pp. 32–64, 1995.

[24] Neissser, U., *Cognition and Reality: Principles and Implications of Cognitive Psychology*, W.H. Freeman and Company: San Francisco, CA, 1976.

[25] Brewer, V.E., A decision making construction for complex situations.. *Engineering Management and System Engineering*, Old Dominion University, 2010.

[26] Rescher, N., Meaning and assertability. *Empirical Inquiry*. Rowman & Littlefield: Totowa, NJ, 1982.

[27] Honderich, T., *The Oxford Companion to Philosophy*, Oxford University Press: New York, 1995.

Techniques for multifractal spectrum estimation in financial time series

P. Jizba & J. Korbel
Department of Physics, Faculty of Nuclear Sciences and Physical Engineering, Czech Technical University in Prague, Czech Republic

Abstract

We show that a multifractal analysis offers a new and potentially promising avenue for quantifying the complexity of various time series. In particular, we compare the most common techniques used for multifractal scaling exponents' estimation. This is done from both a theoretical and phenomenological point of view. In our discussion we specifically focus on methods based on estimation of Rényi entropy, which provide a powerful tool especially in the presence of heavy-tailed data. As a testbed for the applicability of above multifractal methods we use various real financial datasets, including both daily and high-frequency data.
Keywords: multifractal spectrum, Rényi entropy, time series.

1 Introduction

During the past few decades, the concepts of scaling and self-similarity have become common fare in various scientific branches, including dynamical systems [1], biological systems [2], quantum field theory [3] and sociological and economical systems. Global scaling is a key concept, for example, in the theory of critical phenomena and renormalization group, while a global self-similarity is the cornerstone of fractal geometry. Nevertheless, in real systems, such as financial markets, one typically observes local scaling and local self-similarity rules rather than the global ones. A theoretical handle that can successfully deal with such systems is provided by multifractal analysis. This paradigm is based on the assumption that the distribution of local scaling rules possesses also its scaling rule with characteristic scaling exponent called multifractal spectrum. The theory of multifractals has been deeply studied, for

WIT Transactions on State of the Art in Science and Engineering, Vol 90, © 2016 WIT Press
www.witpress.com, ISSN 1755-8336 (on-line)
doi:10.2495/978-1-78466-155-7/009

example, [4–7]. In this connection, there have been developed various methods for multifractal spectrum estimation of time series, such as methods based on generalized Hurst exponent [8] or wavelet transform [9]. Here we will focus on two most common techniques used for estimation of multifractal scaling exponents, namely Detrended fluctuation analysis [10, 11] and the Rényi-entropy-based Diffusion entropy analysis [12–14]. We compare both methods from the theoretical point of view and discuss their applicability in time series analysis. To illustrate theoretical results, we apply both methods to the examples of real financial time series collected on daily and minute basis.

2 Multifractal analysis

Let us have a time series $\{x_i\}_{i=1}^{N}$ measured on the specific time lag s, (e.g. minute basis or daily basis). We group all points into distinct regions K_j, and the probability of occurrence in jth region is given by $p_j = \lim_{N} \infty N_j/N$, where N_j is number of points in K_j. We consider that each probability scales with some characteristic exponent, so $p_j \propto s^{\alpha j}$ and the distribution of scaling exponents is considered in form $\rho(\alpha, s) \propto s^{-f(\alpha)}$. The scaling exponent $f(\alpha)$ is called *multifractal spectrum* and is nothing else than fractal dimension of subset with scaling exponent α. More details can be found in Refs. [7, 15]. Alternatively, one can obtain a related representation of characteristic scaling exponents from so-called partition function $Z(q,s) = \sum_j p_j^q \propto s^{\tau(q)}$. Two comments are now in order. First, the relation between multifractal spectrum and scaling exponent of partition function is given by the Legendre transform $\tau(q) = \max_\alpha [q\alpha - f(\alpha)]$. Second, the exponent $\tau(q)$ is closely related to Rényi entropy $S_q(s) = \frac{1}{q-1}\ln\sum_j p_j^q \propto s^{D(q)}$. Eventually, $\tau(q) = D(q)/(q-1)$. Scaling exponent $D(q)$ is known as a generalized dimension [15]. Rényi entropy was originally formulated by Hungarian mathematician Alfréd Rényi [16] in mid 70 and further developed in number of works (see, e.g. Ref. [17] and citations therein). It represents a one-parametric generalization of Shannon entropy known from information theory and thermodynamics. Because of properties such as observability [18] or relation to multifractals, it has found numerous applications in thermodynamics [17], econophysics [19] and quantum mechanics [20]. In the following section, we compare methods based on estimation of moments of probability distribution with the approach based on Rényi entropy.

3 Estimation of multifractal scaling exponents

The first attempt to describe the scaling exponents in time series was done by Hurst *et al.* [21]. Important upshot of his effort was formulation of the so-called Hurst scaling exponent. The Hurst exponent is calculated as $\langle|\Delta x(t)|\rangle \propto \Delta t^H$ and represents a measure of long-memory and persistence. Nowadays, there are many techniques that estimate Hurst exponent, let us just mention the Rescaled range analysis [22]. Probably the most popular method called Detrended fluctuation analysis (DFA) was originally introduced in Refs. [10, 11, 23]. The key object of the method is the local *fluctuation function* $f(\mathrm{v}, s)$, which sums up fluctuations from a local trend. The global fluctuation function is then calculated as a generalized mean of local fluctuations $F(q,s)=\left(\frac{1}{N_s}\sum_{v}^{N_s} f(v,s)^q\right)^{1/q}$. The fluctuation function scales as $F(q, s) \propto s^{h(q)}$. In some particular cases (e.g. stationarity, positivity), it can be shown that the scaling exponent $h(q)$ is related to conventional multifractal analysis as $\tau(q)= qh(q) - 1$.

An alternative approach to estimate the multifractal scaling exponents can be reached through estimation of Rényi entropy and its scaling exponent. This method is called Diffusion entropy analysis (DEA) and was introduced in Refs. [12, 13] and further discussed in [14]. Unlike other methods based on estimation of moments, this approach can successfully deal with distributions with heavy-tails. As an example, let us have a distribution that is self-similar with the scaling form $p(x,t)=\frac{1}{t_s}F\left(\frac{x}{t_s}\right)$. Then the Shannon entropy ($q \to 1$) is equal to $S_1(t) = A+\delta \ln t$. In case of multiple scaling exponents, these can be revealed by the whole class of Rényi entropies and we obtain $S_q(t)= B_q + \delta(q) \ln t$. At this point, we should stress that proper estimation of Rényi entropy for all values of q is crucial in this method. In practice, this boils down to estimation of probability distribution that is constructed via the so-called *Fluctuation collection algorithm* [12, 14]. To do so, it is necessary to calculate the optimal width of histogram bins so that the interpolating histograms faithfully approximate the underlying distribution. There exist several approaches for such an estimation. We can mention, for instance, the classic rule of Sturges (estimating the optimal number of bins as $1 + \log_2 N$), Scott's rule [24] or Freedman–Diaconis' rule [25] (both estimate the optimal bin widths as being proportional to $N^{-1/3}$). Quite recently, a generalization of the above procedure to the case of Rényi entropy was presented in Ref. [14]. In particular, this provided an estimation of histogram powers $p^{\wedge q}$ for several time scales that is necessary to compute the corresponding scaling exponents.

By comparing the two methods, we can say that the best approach, when examining data, is to combine both of them and depict both spectra $f(\alpha)$ and $\delta(q)$. This is because the transformation of the exponents works, strictly speaking, only in the limit $s \to 0$, which is often replaced by the assumption of exact scaling which is then implemented into the linear regression procedure. Unfortunately, the exact scaling premise is not usually fulfilled in case of real time series. Thus, the knowledge of both spectra brings more complex picture of the multifractal nature of the series. In general, we can say that DFA is better, when the estimation of moments is not problematic, while DEA is more fruitful in case of power law distributions and heavy tails. In the next section, we compare both spectra and time dependence of the Hurst exponent for various real financial series recorded for both daily and high-frequency ticks.

4 Applications of multifractal analysis to financial data

Here we apply the multifractal estimation procedures discussed above to the real financial time series. Financial markets are for this task an ideal testbed since they represent open, non-linear and highly structured complex systems with lots of unexpected and unpredictable phenomena (including sudden jumps, market sentiment, long-memory effects, etc). This, in turn, brings about a non-trivial multifractal structure of market prices reflected in ensuing time sequences. We test the applicability of the aforementioned multifractal techniques on several examples of market time series, particularly on stock index Nikkei 225 (index of Tokyo stock exchange), ASE Composite index (main index of Athens stock exchange), IBM stock and VIX index (implied volatility of S&P 500 options index) recorded on high-frequency and daily basis. Minute time series were recorded during year 2013 and have approximately 10^5 tick points, daily data are from the last 10–20 years (depending on the particular series) and have 5,000–10,000 records. Figs 1 and 2 depict results of the multifractal analysis of all

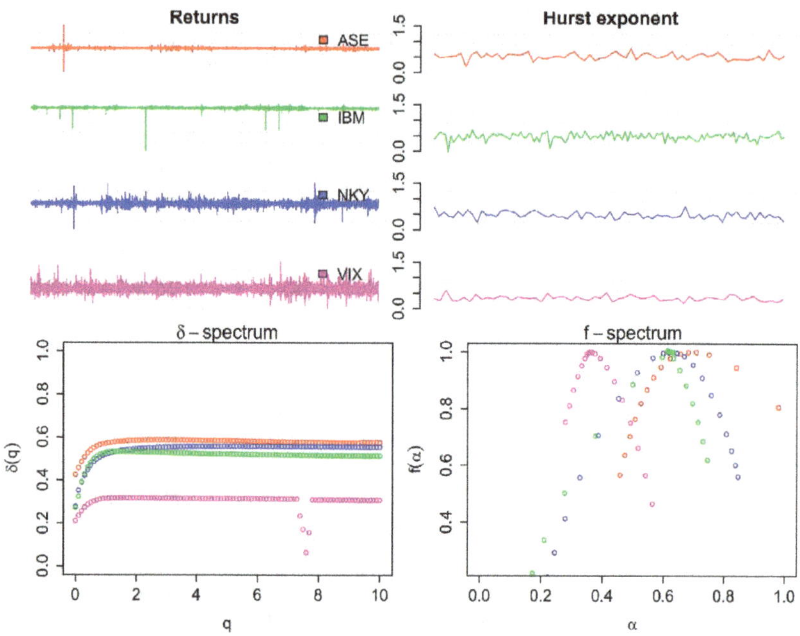

Figure 1: Multifractal analysis of daily data. We can observe that the multifractal exponents of the series VIX are different from the other series. This is caused by a volatile nature of the index, resulting also a discontinuity in the δ-spectrum. Note also that the Hurst exponent of VIX series is noticeably lower than in other series.

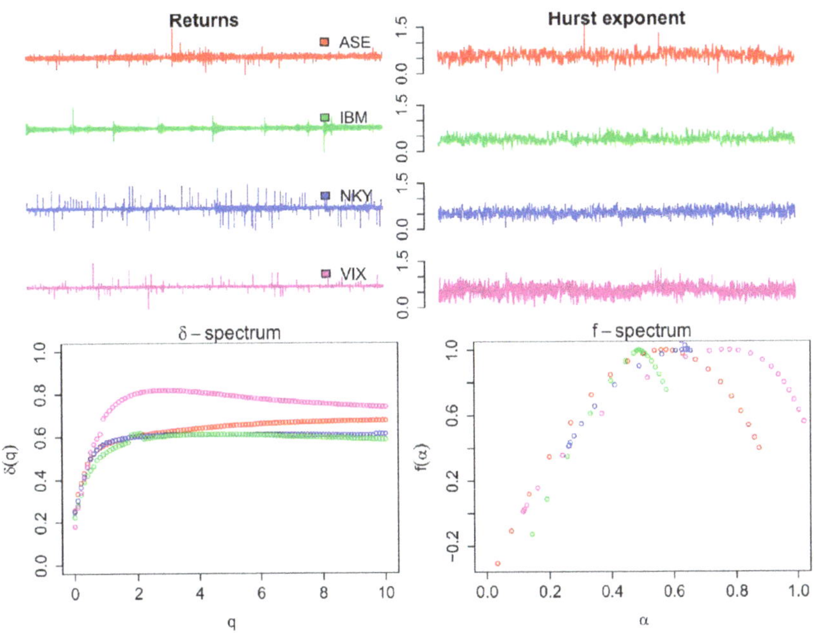

Figure 2: Multifractal analysis of high-frequency financial data. Depicted data are highly non-liquid and exhibit a power-law behavior. This is reflected in discontinuities in both spectra, mainly in the f-spectrum of NKY series.

mentioned series for both characteristic time lags. We can observe that in case of non-liquid series or anomalous scaling, both methods exhibit certain discontinuities in the spectra. This is caused by the fact that estimation of any quantity based on underlying distribution with heavy tails is usually technically difficult and one needs to make very precise calculations. In case of daily series, we observe that the spectra describe mainly the general characteristic scaling exponents, usually close to 0.5 (white noise or Wiener process). This can be attributed to the fact that the data are liquid and correlations of the series decay much faster (order of minutes). On the contrary, minute data have much richer structure of scaling exponents, which is apparent mainly from the δ -spectrum. These two figures show that combination of several multifractal methods is desirable and allows we for a more complete theoretical picture of the observed time series.

5 Conclusions

The multifractal analysis is an important diagnostic tool that allows to revealing a rich and often intricate scaling structure in variety of complex dynamical systems. In this paper, we have compared several techniques for estimation a of multifractal scaling exponents. Notably, we have discussed main theoretical aspects of two currently popular methods, namely detrended fluctuation analysis and diffusion entropy analysis and illustrated their utility in the real-life examples of financial time series. The optimal way is to combine several methods based on different approaches for multifractal spectrum estimation. This can eliminate the numerical and computational artifacts in multifractal spectra that are usually caused by insufficient precision in estimations or by complex nature of underlying data. Nevertheless, such data are usually the most interesting to investigate. In case of financial series, we have shown that different time scales have usually different characteristic scaling exponents. This is because the rough coarse-graining can suppress (and often does!) some non-trivial intermediate dynamics alongside with their scaling exponents and hence the effective coarse-grained spectrum is structurally poorer.

Acknowledgements

Authors want to thank to Xaver Sailer of Nomura Ltd, who provided us with financial data. The work was supported by the Grant Agency of the CTU in Prague, grant No. SGS13/217/OHK4/3T/14 and the GACR grant No. GA14-07983S.

References

[1] Bar-Yam, Y., *Dynamics of Complex Systems*, volume 213. Addison-Wesley Reading, MA, 1997.

[2] Lovejoy, S. & Schertzer, D., *The weather and Climate: Emergent Laws and Multifractal Cascades*. Cambridge University Press, 2013.

[3] Kleinert, H., *Path Integrals in Quantum Mechanics, Statistics, Polymer Physics and Financial Markets, 4th edn.* World Scientific, 2009.

[4] Mandelbrot, B., Self-affine fractals and fractal dimension. *Physica Scripta*, **32**, pp. 257–260, 1985.

[5] Mandelbrot, B., *Multifractals and 1/f Noise: Wild Self-Affinity in Physics (1963–1976)*. Springer, 1999.

[6] Stanley, H.E. & Meakin, P., Multifractal phenomena in physics and chemistry. *Nature*, **335(29)**, p. 405, 1988.

[7] Harte, D., *Multifractals: Theory and Application*. Chapmann & Hall/CRC, 2001.

[8] Morales, R., Matteo, T.D., Gramatica, R. & Aste, T., Dynamical generalized hurst exponent as a tool to monitor unstable periods in financial time series. *Phys A*, **391(11)**, p. 3180, 2012.

[9] Muzy, J.F., Bacry, E. & Arneodo, A., Multifractal formalism for fractal signals: The structure-function approach versus the wavelet-transform modulus-maxima method. *Phys Rev E*, **47(2)**, p. 875, 1993.

[10] Peng, C.K., Buldyrev, S., Havlin, S., Simons, M., Stanley, H. & Goldbergerz, A., Mosaic Organization of DNA Nucleotides. *Phys Rev E*, **49(2)**, pp. 1685–1689, 1994.

[11] Kantelhardt, W., Zschiegner, S., Koscielny-Bunde, E., Havlin, S., Bunde, A. & Stanley, H.E., Multifractal detrended fluctuation analysis of nonstationary time series. *Phys A*, **316**, p. 87, 2002.

[12] Scafetta, N. & Grigolini, P., Scaling detection in time series: diffusion entropy analysis. *Phys Rev E*, **66(3)**, p. 036130, 2002.

[13] Huang, J. et al., Multifractal diffusion entropy analysis on stock volatility in financial markets. *Phys A*, **391(22)**, p. 5739, 2012.

[14] Jizba, P. & Korbel, J., Multifractal diffusion entropy analysis: Optimal bin width of probability histograms. *Phys A*, **413**, pp. 438–458, 2014.

[15] Hentschel, H. & Procaccia, I., The infinite number of generalized dimensions of fractals and strange attractors. *Phys D*, **8(3)**, pp. 435–444, 1983.

[16] Rényi, A., *Selected papers of Alfréd Rényi* sv 2, Akadémiai Kiado,1976.

[17] Jizba, P. & Arimitsu, T., The world according to Rényi: thermodynamics of multifractal systems. *Ann Phys*, 312(1), p. 17, 2004.

[18] Jizba, P. & Arimitsu, T., Observability of rényi's entropy. *Phys Rev E*, 2004.

[19] Jizba, P., Kleinert, H. & Shefaat, M., Rényi information transfer between financial time series. *Phys A*, **391(10)**, p. 2971, 2012.

[20] Jizba, P., Dunningham, J.A. & Joo, J., Role of information theoretic uncertainty relations in quantum theory. *Ann Phys*, **355**, pp. 87–115, 2015.

[21] Hurst, H., Black, R. & Simaika, Y., *Long-term Storage: an Experimental Study*. Constable, London, 1965.

[22] Gilmore, M., Yu, C., Rhodes, T. & Peebles, W., Investigation of rescaled range analysis, the hurst exponent, and long-time correlations in plasma turbulence. *Phys Plasmas*, **9**, 2002.

[23] Koscielny-Bunde, E., Kantelhardt, J., Braun, P., Bunde, A. & Havlin, S., Long-term persistence and multifractality of river runoff records: Detrended fluctuation studies. *J Hydrol*, **322(1–4)**, pp. 120–137, 2006.

[24] Scott, D.W., *Multivariate Density Estimation: Theory, Practice and Visualisation*. John Wiley and Sons, Inc., 1992.

[25] Freedman, D. & Diaconis, P., On the histogram as a density estimator. *Zeitschrift für Wahrscheinlichkeitstheorie und Verwandte Gebiete*, **57(4)**, p. 453, 1981.

Section 2
Applications in business and industry

Addressing supply-chain complexity using closed-loop simulation-based exercises

C. W. Axelrod
Delta Risk LLC, USA

Abstract

From news reports about companies attempting to reduce the impact of compromised supply chains, due to natural disasters, accidents or targeted attacks, or trying to avoid specific products or ingredients banned on moral grounds, it is apparent that many organizations have only rudimentary knowledge of the provenance of software, hardware, and other supplied items. Reasons for this situation include the difficulty and effort required to:

- build and maintain complete and accurate databases;
- obtain information on subcontractors down to the required level of detail;
- review, monitor and test products to ensure that they are genuine;
- encourage eradication of deficiencies, weaknesses, and vulnerabilities;
- ensure that changes are identified, reported, analyzed, and addressed;
- identify commonalities and common points of failure;
- introduce resiliency, redundancy, and backup within the supply chain;
- develop methods to simulate infrastructures, transactions, etc.; and
- bring together competitors to collaborate in exercising various scenarios.

Thus, the question arises as to how to resolve these issues in an accurate, efficient, and cost-effective manner. Answering this question is our goal.

┬ upply-chain models are generally substantially more intricate than the model developed for the US equities marketplace. However, the same approach works for developing and operating any complex industry-wide and sector-wide systems with many participants who want to keep proprietary information confidential but need to share information to facilitate a rich exercise experience

WIT Transactions on State of the Art in Science and Engineering, Vol 90, © 2016 WIT Press
www.witpress.com, ISSN 1755-8336 (on-line)
doi:10.2495/978-1-78466-155-7/010

for learning, training, and testing a variety of realistic scenarios. This paper describes a process for implementing such simulation-based exercises.
Keywords: closed-loop tabletop exercises, commonalities, complexity, complicatedness, counterfeiting, resiliency, supply chain, tampering, transaction-level simulation models, vendor management.

1 Introduction

Prior to the industrial revolution, supply chains consisted of local suppliers providing limited ingredients to artisans, such as flour to bakers, leather to cobblers, iron to blacksmiths, etc. Over time, products and services became more sophisticated, requiring harder-to-come-by components from larger specialty producers, who benefited from economies of scale. At that time, supply-chain - vendors and outsourcing partners were usually well known to customer organizations. Suppliers were mostly domestic, if not local. In these circumstances, they could be audited on-site, with little or no advanced notice. However, suppliers commonly imported raw materials, such as spices, and intermediate products, such as fabrics, from distant lands. This made for additional risk since early forms of transportation, such as sail ships, were often hazardous and, as a result, schedules were uncertain.

Today, supply chains are commonly distributed globally with suppliers thousands of miles from their customers. Also, many components incorporate elaborate software systems and intricate electronic components, which may have been designed in one country, manufactured in another, and assembled in yet a third. As a result, the management of modern supply chains has become increasingly difficult, requiring ever more highly skilled resources. In this paper, we discuss the issues and suggest ways in which many of the problems can be resolved, or at least mitigated.

2 Simplicity, complicatedness, and complexity

The first question to be addressed here is about the very nature of complex systems, from which we can determine if indeed supply chains can be considered to be complex or merely complicated. The differences among simple, complicated, and complex systems have been discussed extensively in the literature, as in [1–8].

The consensus seems to be that complicated systems are aggregates of many components that must operate together in planned, structured, and predetermined ways. Complex systems, on the other hand, are held to be adaptive in response to continual change and uncertainty. This implies that complicated systems are deterministic, whereas complex systems are probabilistic in nature.

We can see distinctions among simple, complicated, and complex systems in Table 1. For example, complicated systems are synergistic, that is, as simpler systems are combined into more complicated systems, the latter may well exhibit capabilities that are greater than the sum of the capabilities of the individual parts. This is similar to what system engineers call 'systems of systems'.

Table 1: Definitions, descriptions, characteristics, and examples of simple, complicated, and complex systems.

Attributes	Nature of system		
	Simple	Complicated	Complex
Definition [1]	Systems for which the chances of success are high and easily predicted	Systems that follow patterns and about which accurate predictions can be made	Systems where interactions and consequences are difficult to determine
Description [2]	May require some basic technique and terminology, but then have high assurance of success	Not just an assembly of simple components Nature is related to scale Requires coordination and specialists	Cannot be reduced to complicated problems Requirements based on local conditions, interdependencies, and ability to adapt to change Ambiguous and uncertain
Characteristics [3]	Can be replicated easily Expertise not required Products standardized Consistent results	Assurance of success from prior experience Expertise needed Some commonalities Results certain	No assurance of success Expertise not sufficient Each project unique Outcome uncertain
Examples [3, 4]	Following a recipe	Sending a rocket to the moon Building a highway Building an air-traffic control center	Raising a child Managing traffic congestion Directing air traffic

Complex systems, on the other hand, are fraught with uncertainty and cannot be managed in the same way as are complicated systems. They need much more flexible controls and the ability to respond quickly to unanticipated events or changes. To some extent, one can consider complex systems to be complicated systems that have been subjected to perturbing random events, so that their formerly stable operation becomes volatile.

3 Are supply chains complicated or complex?

When it comes to modern supply chains, they are seem to be increasingly hard to manage, although researchers, such as Brody [9], claim that certain technologies, for example, 3-D printing, will result in simpler relationships between customer organizations and local suppliers, with fewer interdependencies. However, there are many economic factors, such as the availability of skilled labor, cheap raw materials, and efficient transportation, as well as environmental restrictions, which limit the attractiveness of local sourcing.

There seems to be agreement that most modern supply chains are not simple. The issue is whether they are complicated, that is, they can be planned, managed, and controlled, or whether they are complex, that is, they are dominated by uncertainty and the need to adapt quickly to changing circumstances.

From the business perspective, senior executives' want to control supply chains completely, which suggests that they see modern supply chains as merely complicated, not complex. Unfortunately, organizations cannot escape uncertainties experienced through their supply chains as suppliers are dispersed globally among countries with varying cultures, infrastructures, legal systems, and weather patterns, and vulnerability to earthquakes, volcanoes, floods, fires, tsunamis, etc. Therefore, we must accept that many supply chains are indeed complex and we must do our best to mitigate risks that such systems experience.

4 Reasons for supply-chain complexity

Deshpande [10] gives five reasons why supply chains are complex. He asserts that a large number of variables, which he calls 'numerousness', makes for complex supply chains. However, complicated systems also have many variables, so that numerousness alone cannot define complex systems. Then, there is 'variety', which refers to variables' distribution patterns, such as - seasonal demand. Again, if such distributions are known, then such a supply chain is complicated rather than complex. However, if the distributions show high variability, then systems should be viewed as complex. He claims that 'interconnections' also determine whether systems are complex. On the other hand, however many interactions there might be among parameters, the system remains complicated as long as the interconnections are well defined and stable over time. If there is variability or uncertainty with respect to interconnections, then systems are considered complex.

'Opacity' means that the 'exact nature of (a) relationship' may not be known with certainty. This is a core characteristic of supply-chain complexity. Opacity is often the main reason why companies get into trouble when there are catastrophic events, such as earthquakes or tsunamis, or when governments outlaw certain materials, such as 'blood diamonds' and North Korean gold.

A major recommendation of this paper is to address opacity through anonymous information sharing facilitated by tabletop exercises incorporating simulation models.

'Dynamic effects' relate to small perturbations that are magnified as they propagate throughout supply chains, in the so-called 'bull whip' effect. This supports the idea that supply chains are complex as, for example, when there is a 'lack of information on the true nature of product demand'.

When opacity and dynamic effects are combined, you have the worst of both worlds with respect to supply-chain complexity. For example, relatively small events, not reported to customer organizations quickly enough, may continue to grow until their effects are so great that they severely disrupt business-as-usual and become newspaper headlines.

5 Building vendor management databases

The main antidote to complexity is timely, accurate, and sufficient information. For supply chains, this translates into gathering, analyzing, and acting upon information obtained directly from suppliers or garnered from other parties. For many organizations, the oversight of suppliers is handled by their vendor management departments in concert with internal or outside lawyers. When operating as they should, the vendor management and the legal departments are responsible for collecting data from both internal and external sources on all suppliers dealing with the organization. It is common to use pro-forma surveys for gathering such information. Each vendor is asked to complete the survey in order to obtain the information needed to approve or turn down the vendor. In some cases, the level of detail may vary depending upon the criticality of the supplier to the customer organization. These questionnaires typically consist of dozens of pages of questions, usually with 'yes', 'no', and 'not applicable' responses to facilitate scoring. The analysts may weight the answers with respect to importance, and the resulting score is the weighted sum or weighted average of the aggregated scores. If the total of the weighted scores falls below a particular 'hurdle' number, the vendor may be rejected outright or might be asked for further details related to their initial responses. For example, if a vendor responds that they have a set of security policies, they may be asked to supply evidence in the form of the policies or an audited review that has evaluated the policies and confirmed their existence and appropriateness. The additional information augments the data already collected and all the information goes into a more rigorous evaluation. A process of this type has been formalized in, for example, the Santa Fe Group's Shared Assessment Program (SAP) described in [11], details of which can be reviewed by the reader at http://santa-fe-group.com/capabilities/shared-assessments/. The SAP program grew out of a project originated by the BITS division of the Financial Services Roundtable, which represents major US financial institutions [12].

In many cases, organizations send out vendor surveys annually, or more frequently if the vendor is very critical to the customer organization or if something happens that impacts the products and services being supplied and received. Timing and periodicity, level of detail of information requested, and the like will vary depending upon the criticality of the vendor to the business or as a result of known significant changes in vendors' circumstances, due to, for example, mergers and acquisitions, changes in physical and financial health, business strategies, and the like. If a vendor is important enough to the business, an on-site review might be warranted. If a vendor supplying a European or US organization and/or its subsidiaries works out of Asia, say, it will probably be more cost-effective to employ a trusted local auditing firm for on-site reviews and to report findings back, rather than transporting internal staff to a distant site.

Whichever approach is selected, and wherever critical vendors are located, the main result of all these data gathering effort is a vendor management database containing financial and operational details for each supplier. Also, when non-public personal information, such as data about customer organizations' own or their business partners' customers or employees, is collected and used by service providers, then a whole series of laws and regulations kick in, depending on the location of suppliers and customers. In some cases, countries in which the customer organizations are located determine which legal and regulatory requirements must be applied, even if the supplier is located in a different country. The database serves as a repository of vendor information which informs managers about such details as the date of the prior review, previous scores, and the like, making for much improved supply-chain risk management.

Another mechanism that needs to be in place is a notification process through which the database can be updated and, more importantly, it can be determined whether an audit or business review should be initiated. Unless an effective change notification process is in place, it is very likely that the customer organization will not hear about significant changes in suppliers' financial, - organizational, managerial, and operational situations until an adverse event occurs, a change of location is made, or a supplier engages a new subcontractor or outsources certain activities, which brings changes to the fore.

We show, in Table 2, data that need to be collected, monitored, and received as notifications from vendors, vendor-management processes, and material events. We also indicate actions that need to be taken in order to ensure that supply-chain events can be properly managed.

6 Vendor and subcontractor data collection

While it is relatively straightforward, though not easy, to obtain survey information from direct suppliers, since they are motivated to remain as suppliers, it has been found to be much more difficult to obtain information about suppliers' subcontractors and vendors. This is in part because companies often view information about their suppliers to be confidential and proprietary. For example, when surveyed about the sources of meat in their products, companies that did respond (several did not) gave the following reasons for not supplying the requested information [13]:

"...we do not normally discuss our sourcing strategies"

"(We do not) provide (vendor) names ... as this is proprietary"

"We don't disclose supplier names for competitive reasons"

"We do not disclose any vendor or supplier information..."

Table 2: Vendor management database actions.

Category	Periodicity	Type of data	Actions
Vendor	Annual Semi-annual On-demand Triggered	Financial Operational Physical and information security Information security Legal and regulatory Locations and jurisdictions Business continuity/disaster recovery	Accept/reject vendor Request additional information Request audit review reports Request security review reports Obtain any certifications Request on-site visit(s) by company or third party
Process	Continuous	Efficiency Effectiveness Accuracy Risk Resiliency	Monitor process to ensure that is in proper working condition Perform third-party reviews Re-evaluate scope of data and risk criteria when bad situations arise
Events	Random	Natural disasters Political situations Local fires, floods, etc. Business events, e.g. takeovers, mergers, relocations Bankruptcies Other business cessation reasons	Assess impact of event Review contingency plans Determine short/long-term options Select appropriate response Initiate business continuity plans Initiate disaster/recovery plans Look for other suppliers/customers

Although it is likely that companies are more willing to disclose supplier information on a confidential basis, where it will not be made public, than to the press, they are still very reluctant to reveal such information even on a confidential and anonymous basis. Also, it is often difficult to convince subcontractors that there is value to them in providing requested information.

A survey conducted on behalf of the US financial services industry, described in greater detail in [14], found that, while many institutions were willing to describe their vendor-management processes and procedures, far fewer responded when it came to actually listing all their suppliers. This appeared to be due to firms' not having readily accessible databases that could have provided such a list. Ironically, many organizations put together such lists as part of their Y2K data-collection efforts, but following Y2K they generally failed to keep these lists up to date. Also, particularly in large organizations, this type of information is scattered across many divisions and departments and does not always exist in a central repository. Even when these lists were developed and maintained, they seldom, if ever, contained information about subcontractors and vendors to suppliers.

Given such a response, one can anticipate considerable resistance from direct suppliers if they were asked to list subcontractors and second-level suppliers. And yet a complete picture is not attainable unless everyone agrees to participate and share information. As discussed later, responses might be better if companies were able to submit information in an anonymous, yet authenticated, manner.

7 Tampering and counterfeiting

In some industries, such as the pharmaceuticals industry, fake products are a huge issue. With physical products, in particular, there are a number of ways to identify the genuine article. However, it is very difficult to know whether there has been some form of tampering of software. There are usually no traces left when software is copied, as would be the case with physical theft. On the other hand, there exist methods to ensure that software cannot be accessed or copied, preserving the original asset and preventing tampering or theft.

Furthermore, extremely complicated equipment, comprising myriads of integrated computer chips that are designed and fabricated using specialized software, can be tampered with at the design, development, manufacturing, and testing phases as well as during maintenance.

Knowledge of the provenance of software and equipment, achieved by tracing through each step in the supply chain, can go a long way towards determining whether specific products should be suspected of having been subjected to tampering or counterfeiting. However, in many situations there is no substitute for full laboratory testing.

8 Checking for deficiencies

It is incumbent upon vendors and subcontractors to check their products for software vulnerabilities and weaknesses, as well as for mechanical deficiencies. However, it is up to customer organizations to perform due-diligence reviews to assure themselves that proper care has been applied to the assurance of third-party products and services. One means of achieving this is by incorporating one's own organization's standards, processes, and procedures into contracts with third parties. A particularly strongly worded contract for the acquisition of third-party software was developed by Pelgrin *et al.* [15] in the SANS Application Security Procurement Language document.

When a large customer has relatively small suppliers, that is, suppliers provide the bulk of their output to one particular customer organization, the latter will normally have significant leverage over such suppliers. Under these circumstances, customer organizations can often dictate strong terms in service or product contracts and ensure that they will be permitted to monitor compliance with those terms and conditions. Similarly, large suppliers are able to have customers sign purchase agreements that significantly reduce supplier liability. This balance of power is discussed by Axelrod [16]. However, for the majority of supplier–customer relationships, organizations can neither get agreement on such stringent terms nor can they effectively enforce requests to be informed immediately when adverse events occur. The following section suggests an approach to alleviate this issue.

9 Change notifications

One of the greatest challenges in supply-chain management is ensuring notification when material changes take place within a particular supply chain. Many suppliers are understandably concerned about revealing proprietary information that might give competitors an edge or bring law suits down upon them. Consequently, they are often reluctant to provide information about incidents that might lead to delays, compromises, or cessation of supply.

Because of this reluctance, we need some method of reducing risks of suppliers making critical information about incidents available to customers on a timely basis, so that the latter can take necessary actions. One suggested means of achieving this sharing of crucial information is for all entities in the supply chain to provide information to a central monitoring body, which then reports significant events to customers. However, if preferred, incident reports can be sent to a trusted third party for distribution to various interested parties, including customers, suppliers, regulators, law enforcement, government agencies (such as emergency and disaster recovery services), and the public at large. The template for such an information-sharing system, in which identities of submitters can remain anonymous and content can be scrubbed to de-identify any attributable information, already exists in the form of industry ISACs (Information Sharing and Analysis Centres). The role of ISACs includes making sure that information comes from authentic sources and adding intelligence obtained from other sources, as shown in Fig. 1.

Of course, suppliers and customers need to be identified if action is required by business partners, regulators, law enforcement, and even the military. Protocols need to be developed for such situations. A key success factor is a measure of protection so that, for example, proprietary information is not leaked to the public and parties involved have time to prepare a response.

This flowchart can be used as the basis of a computer simulation model, to represent how one might obtain data and report them for complex supply chains, and thereby enable the various entities to respond and resolve issues resulting from incidents. In this manner, the simulation model becomes the basis for closed-loop transaction-level tabletop exercises, as discussed below.

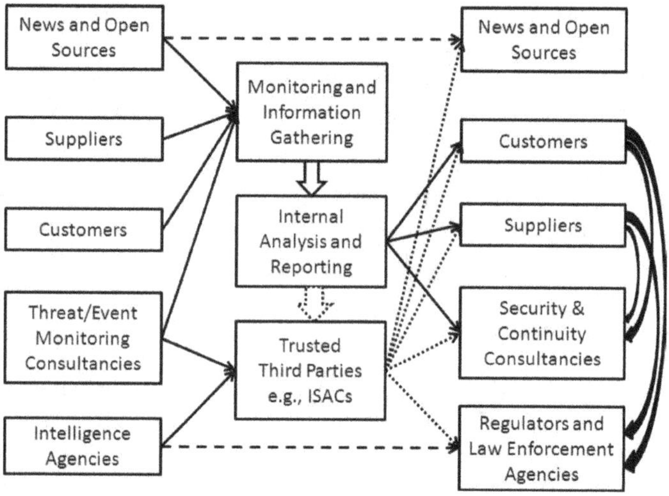

Figure 1: Supply-chain information gathering, analysis, and reporting.

10 Common points of failure

Modern economic systems demand the most efficient production and distribution methods as companies compete on a global scale. This approach favors the trend towards large-scale suppliers and distributors in order to achieve requisite economies of scale. However, this move to fewer and larger entities, along with the just-in-time inventory philosophy to reduce inventory levels and their respective costs, results in more common points of failure, as described by Axelrod [17]. For example, huge manufacturers, such as Foxconn, build electronics equipment for many retail brand names, and suppliers of specialized components provide their products to many different manufacturers and retailers, such as automobile makers.

11 Resiliency versus economics

As mentioned above, there is often a possible tradeoff between supply-chain resiliency and costs. The former can be achieved by using a bigger number of suppliers and carrying larger inventories, as described in [17]. However, the pendulum is swinging in the other direction. That being the case, the means of risk mitigation is to have a store of data that can be used in preparation for incidents and in real-time when incidents occur. Because these data are difficult and expensive to come by, and, once they have been acquired, require deep understanding and constant updating, the only feasible approaches for handling such vast and rapidly changing data are the use of big-data analytics and computer simulation models, accompanied by frequent reviews and tests to ensure that all the information is current. We will not be addressing the big-data analytical methods in this paper. Suffice to say that the resulting computer simulation will be much superior if relationships among entities, transactions, and events have been established through a combination of analytical results and subject-matter expertise.

12 Simulation models and exercises

There are a variety of ways in which supply-chain simulation can be approached. The choices are among emphasizing infrastructure or transactions, restricting the model to specific marketplaces or expanding it for broad coverage, determining the level of detail, choosing between open-loop or closed-loop models, combining in-person players and virtual players, and the like. Decisions as to the structure and functionality of the model depend mostly on the purpose of the model as well as the cost and difficulty of constructing and operating it. As the DECIDE-FS® model for financial services demonstrated, attaining an accurate, realistic model can be a multi-million-dollar effort extending over many years.

A supply-chain model, even if limited to a single industry, such as automakers or the pharmaceutical industry, is even more complex than that of a single industry, such as financial services, and so it can be expected to be significantly more costly and difficult to build a good representation of supply chains as cooperation among participants will likely be less. Nevertheless, the pay-back can be huge.

Figure 2 illustrates the structure of a closed-loop transaction-level simulation model and how it fits within an exercise environment. This particular diagram shows a combination of real-world and virtual players, which has been successfully used in several tabletop exercises for the US securities industry. More detailed descriptions of such exercises, as they relate to financial services and supply chains are given in Axelrod and Schmidt [18] and Axelrod [19] respectively. The characteristic to note is the closed-loop feedback arrow, by means of which lessons learned within and between exercises are fed back into the model to serve to improve future exercises. A surprising number of other forms of exercise do not have this type of feedback so that each effort is essentially based on little or no past experience from prior exercises.

13 Critical success factors

In addition to getting the model right, to be effective, exercises must not only be well coordinated but they should also exhibit the following properties:

- The right mix of organizations, representing each of the player groups, should participate in exercises, either on-site or remotely.
- Specific individuals from each organization should be carefully selected with subject-matter experts and decision-makers at hand throughout exercises.
- Those running exercises should be very familiar with both the operation of simulation models and how to conduct such exercises.
- Exercises should take place periodically, e.g. twice a year, or when a particular event or situation arises, e.g. the possibility of a pandemic.

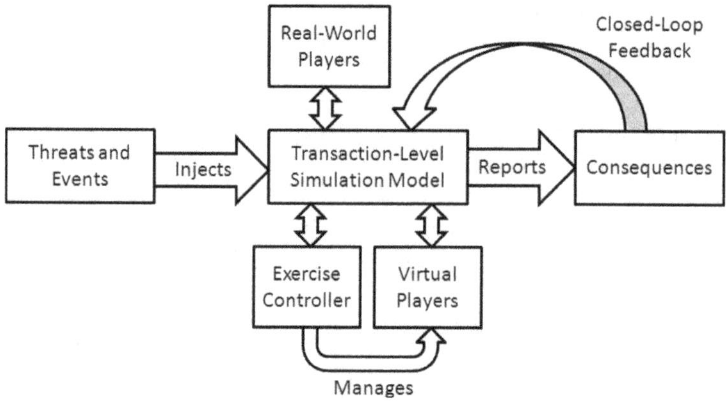

Figure 2: Closed-loop exercise using transaction-level simulation model.

14 Conclusions

Supply chains are becoming increasingly complex and costly to manage, if indeed they can be fully managed. Major incidents have far-reaching effects on the ability of supply chains to operate in a well-controlled manner, and many organizations have neither the tools nor the management support to engage in expensive model building and exercises. Furthermore, sharing accurate, timely information is needed to understand weaknesses in supply chains and to establish greater resiliency of products and services.

Since interdependencies within supply chains are becoming more complex as new countries climb the industrialization ladder, the need for effective tools, greater communication, and more collaboration among business partners and competitors is becoming ever-more crucial. Whereas it would be ideal if supply chains could be simplified, they seem to be heading in the other direction, so that it is incumbent upon us to develop means of dealing with such complexities in order to manage and control supply chains which are increasingly critical to continued economic progress.

References

[1] Lelong, A., Complicated systems vs. complex systems, *Global Supply Chain News*, Global Supply Chain Group, December 2013.

[2] Glouberman, S. & Zimmerman, B., Complicated and complex systems: what would successful reform of medicare look like? *Discussion Paper No. 8,* Commission on the Future of Health Care in Canada, July 2002.

[3] Allen, W., Complicated or complex – knowing the difference is important, Sparks for Change, March 2013.

[4] Kamensky, J.M., *Managing the Complicated vs. the Complex*, IBM Center for the Business of Government: Washington, DC, Fall/Winter 2011.

[5] Sargut, G. & McGrath, R., Learning to live with complexity, *Harvard Business Review*, **89(9)**, pp. 68–76, 2011.

[6] Snyder, S., The simple, the complicated, and the complex: educational reform through the lens of complexity theory, *OECD Education Working Papers No. 96*, December 2013.

[7] Strauss, V. & Cuban, L., The difference between "complex" and "complicated" – and why it matters in school reform, The Washington Post, August 8, 2014.

[8] Waldrop, M.M., *Complexity: the Emerging Science At the Edge of Order and Chaos*, Touchstone, Simon & Schuster: New York, NY, 1992.

[9] Brody, P., Today's Complex Global Supply Chains are Poised to be Dismantled, *Gigaom*, July 2, 2013.

[10] Deshpande, B.R., Top 5 reasons for supply chain complexity – measuring and monitoring complexity to generate early warnings, *Ontonix*, July 2010.

[11] Shared Assessments Program, Agreed Upon Procedures (AUP); Standard Information Gathering Questionnaire (SIG), The Santa Fe Group, 2014.

[12] BITS Framework: Managing Technology Risk for Information Technology (IT) Service Provider Relationships, BITS, October 2001.

[13] Shanker, D., 11 Food companies that won't tell you where their meat comes from, *BuzzFeed Life*, April 10, 2014.

[14] Axelrod, C.W., Malware, 'weakware', and the security of software supply chains, *CrossTalk Journal*, March/April, 2014.

[15] Pelgrin, W., Routh, J. & Williams, J., SANS Application Security Procurement Language, SANS, 2009.

[16] Axelrod, C.W., *Outsourcing Information Security*, Artech House, Norwood, MA, 2004.

[17] Axelrod, C.W., Risks of unrecognized commonalities in the information technology supply chain, *IEEE International HST Conference*, Waltham, MA, 2010.

[18] Axelrod, C.W. & Schmidt, R., A successful transaction level simulation model of the US securities marketplace, *IEEE International HST Conference*, Waltham, MA, 2012.

[19] Axelrod, C.W., Using transaction-level simulation to prepare for and recover from supply-chain disasters, *IEEE International HST Conference*, Waltham, MA, 2013.

Price competition strategy of internet platforms

L. Zhuang
School of Economics and Management, Southeast University, China

Abstract

This paper generalizes the economic features of internet platforms and analyses their competitive strategies with the method of insulating pricing. The analysis focuses on the quantity expectations game of platforms and their price expectations game. The function of network externalities on the internet platforms has an obvious feature of two stages, so their competitive strategies have to be divided into two distinct phases. Before the critical mass stage, companies need to pursue a rapid increase of subscribers in order to survive. After the critical capacity, differentiation strategy can be implemented by platform companies which seek maximum profit by the action of network externalities.
Keywords: competitive strategy, insulating pricing, internet platform, non-cooperative game.

1 Introduction

In the internet era of big data, various internet platforms stake to fierce competition, which leads to a problem about how the internet platforms compete. Compared with the traditional competitive strategies, competitive strategy of internet platforms hides the key factor to success [1, 2]. This article attempts to explain competitive strategy of internet platforms from a perspective of the network economic, which provides theoretical support and practical guidance for competition on the internet platforms [3].

Internet platforms are the product of network economic development, so they have a lot of different characteristics compared with traditional companies [4]. First, platforms have significant network externalities [5], which are generally divided into membership and usage externalities [6, 7]. In addition to the analysis of network effects of platforms [8], more literature discusses pricing strategy of platforms from membership externalities and usage externalities [9]. Weyl [10]

WIT Transactions on State of the Art in Science and Engineering, Vol 90, © 2016 WIT Press
www.witpress.com, ISSN 1755-8336 (on-line)
doi:10.2495/978-1-78466-155-7/011

analyses optimal pricing of platforms under different targets with the method of insulating tariffs [11]. A more realistic problem is how platforms participate in the competition [12]. Hence, the paper mainly analyses the competitive strategies of internet platforms through two-sided markets and game theory.

This article is structured as follows: Section 1 introduces the basic features and model assumptions of the internet platforms. Section 2 constructs theoretical models of platforms competition, includes the dynamic model about users' scale expectations and price expectations. Section 3 summarizes the full paper and makes some relevant suggestions.

2 Theoretical model

To analyze competitive strategies of platforms more clearly, this article gives some assumptions:
(1) This analysis focuses on two platforms (*i*, *j*). When an internet platform makes decisions, the other platforms are considered as a whole by it, namely two platforms in the market. Platforms have two sides, *d* side and *s* side.
(2) Only considering cross externalities of platforms, without considering externalities in the same side of platforms. Because the role of network makes marginal cost providing services to users is close to zero.

2.1 Dynamic game model of quantity expectations

First, we analyze how bilateral platforms take price competition with the methods of insulating tariffs and Game of expectation about users' quantity or scale (as amended Cournot model). Internet platforms have cross network externalities. The price of one side by platform charged is based on the (expected) quantity of users on the other size [13].

Therefore, the pricing function of competitive platforms is as follows:

$$\begin{cases} p_i^d(n_i^s) = a_i^d n_i^s + h_i n_j^{de} \\ p_i^s(n_i^d) = a_i^a n_i^d + h_i n_j^{se} \end{cases}; \begin{cases} p_j^d(n_i^s) = a_j^d n_j^s + h_j n_i^{de} \\ p_j^s(n_j^d) = a_j^s n_j^d + h_j n_i^{se} \end{cases} \tag{1}$$

where, h_i n_n^{de} is the competitors' market share of a platform and h_i is the adjustment factor which can be either positive or negative, but it is usually negative in the model about expected quantity. It means that the price of the platform charged and user scale of the other side are negatively correlated.

Profit function of platforms can be transformed into:

$$\begin{cases} \pi_i = (a_i^d n_i^s + h_i n_j^{de} - c_i^d) n_i^d + (a_i^s n_i^d + h_i n_j^{se} - c_i^s) n_i^s - C_i \\ \pi_j = (a_j^d n_j^s + h_j n_i^{de} - c_j^d) n_j^d + (a_j^s n_j^d + h_j n_i^{se} - c_j^s) n_i^s - C_j \end{cases} \tag{2}$$

Proposition 1: In the game about expected quantity, the price of one side set by platforms is determined by cross network externalities and expected adjustment of competitors' market scale. Final equilibrium results are:

$$\begin{cases} n_i^d = \dfrac{c_i^s - h_i n_j^{se}}{a_i^d + a_i^s} \\ n_i^s = \dfrac{c_i^d - h_i n_j^{de}}{a_i^d + a_i^s} \\ n_j^d = \dfrac{c_j^s - h_j n_i^{se}}{a_j^d + a_j^s} \\ n_j^s = \dfrac{c_j^d - h_j n_i^{de}}{a_j^d + a_j^s} \end{cases}; \quad \begin{cases} p_i^d = \dfrac{a_i^d c_i^d + a_i^s h_i n_j^{de}}{a_i^d + a_i^s} \\ p_i^d = \dfrac{a_i^s c_i^s + a_i^d h_i n_j^{se}}{a_i^d + a_i^s} \\ p_j^d = \dfrac{a_j^d c_j^d + a_j^s h_j n_i^{de}}{a_j^d + a_j^s} \\ p_j^s = \dfrac{a_j^s c_j^s + a_j^d h_j n_i^{se}}{a_j^d + a_j^s} \end{cases} \tag{3}$$

From the result, the users' scale of one side on the platform is decided by marginal cost of providing theses service, target users' scale of competitors and cross network externalities.

The price of one side on the platform is depended on cross network externalities, the expected share of competitors and marginal cost.

Then, we analyze the impact of each factor on the platform pricing using the method of control variables:

(1) *Marginal cost.* Assuming that the cross network externalities coefficient and the expected scale of competitors are constant, the price function is:

$$p_i^d = \frac{\overline{a_i^d} c_i^d + \overline{a_i^s}\, \overline{h_i n_j^{de}}}{\overline{a_i^d} + \overline{a_i^s}}.$$

It means that the price set by platforms is a positive correlation with the marginal cost of platforms' services. This is consistent with the law of economics that the higher the cost of these services provided for each customer interaction, the higher the price charged by platforms.

(2) *Expected scale of competitors' other users.* Assuming that cross network externalities coefficient and marginal cost are constant, the price function is:

$$p_i^d = \frac{\overline{a_i^d}\, c_i^d + \overline{a_i^s} h_i n_j^{de}}{\overline{a_i^d} + \overline{a_i^s}} \quad \text{(where } h_i < 0\text{).}$$

It means that the price set by platforms is a negative correlation with the expected scale of users on the other side of other competing platforms. This is consistent with the law of economics that the more the expected scale of users on the other side of other competing platforms, the lower the price charged by platforms.

(3) *Cross network externalities coefficient.* Assuming that the expected scale of competitors and marginal cost are constant, the price function is:

$$p_i^d = \frac{a_i^d \overline{c_i^d} + a_i^s \overline{h_i n_j^{de}}}{a_i^d + a_i^s} \quad \text{(where } h_i < 0\text{)}.$$

When $c_i^d > h_i\, n_j^{de}$ cross network externalities coefficient a_i^d is a positive correlation with the price charged by platforms, and cross network externalities coefficient a_i^s is a negative correlation with the price charged by platforms. When $c_i^d < h_i\, n_j^{de}$ cross network externalities coefficient a_i^d is a negative correlation with the price charged by platforms, and cross network externalities coefficient a_i^s is a positive correlation with the price charged by platforms.

When the scale of platforms exceeds critical capacity, marginal cost of the service provided by platforms is close to zero, $c_i^d = 0$. Then,

$$p_i^d = \frac{a_i^s \overline{h_i n_j^{de}}}{a_i^d + a_i^s},$$

which means that the price is mainly influenced by cross network externalities coefficient. When the scale of one side on the platform i increases gradually and makes $c_i^d > h_i\, n_j^{de}$ the competition of platforms can reach an equilibrium finally.

Thus, internet platforms has an obvious feature of two stages. After the scale of users exceeds critical capacity, platforms can increase the pricing for service based on the size of externalities, and then enter the competitive differentiation stage.

Corollary 1: In repeated games, the equilibrium prices for internet platforms are:

$$\begin{cases} p_i^d = \dfrac{a_i^d c_i^d (a_j^d + a_j^s) + a_i^s h_i c_j^s - h_i h_j c_i^d}{(a_i^d + a_i^s)(a_j^d + a_j^s) - h_i h_j} \\ p_i^s = \dfrac{a_i^s c_i^s (a_j^d + a_j^s) + a_i^d h_i c_j^d - h_i h_j c_i^s}{(a_i^d + a_i^s)(a_j^d + a_j^s) - h_i h_j} \\ p_j^d = \dfrac{a_j^d c_j^d (a_i^d + a_i^s) + a_j^s h_j c_i^s - h_i h_j c_j^d}{(a_i^d + a_i^s)(a_j^d + a_j^s) - h_i h_j} \\ p_j^s = \dfrac{a_j^s c_j^s (a_i^d + a_i^s) + a_j^d h_j c_i^d - h_i h_j c_i^s}{(a_i^d + a_i^s)(a_j^d + a_j^s) - h_i h_j} \end{cases} \qquad (4)$$

So, these competing platforms can reach an equilibrium state under the condition that platforms set a price based on the coefficient of interactive value in the game of quantity expectations.

To simplify the analysis, the interaction coefficient and costs of the same platform do not distinguish, namely $a_i^d = a_i^s = a_i$, $c_i^d = c_i^s = c_i$, $c_j^d = c_j^s = c_j$. Thus, the price function can be simplified as

$$p_i = \frac{2a_i c_i \overline{a_j} + a_i h_i \overline{c_j} - h_i \overline{h_j} c_i}{4a_i \overline{a_j} - h_i \overline{h_j}} \qquad (5)$$

The price on the side set by platforms depends on the interaction coefficient, marginal cost factor and the expected users' share of opponents.

Partial derivative on the price factor is as follows:

$$\begin{cases} \frac{\partial p_i}{\partial a_i} = \frac{(2\overline{a}_j c_i - h_i \overline{c_j})h_i \overline{h_j}}{(4a_i \overline{a}_j - h_i \overline{h_j})^2} \\ \frac{\partial p_i}{\partial c_i} = \frac{2a_i \overline{a}_j - h_i \overline{h_j}}{(4a_i \overline{a}_j - h_i \overline{h_j})} \\ \frac{\partial p_i}{\partial h_i} = \frac{2a_i \overline{a}_j (2a_i \overline{c_j} - c_i \overline{h_j})}{(4a_i \overline{a}_j - h_i \overline{h_j})^2} \end{cases} \tag{6}$$

(1) *Expected scale of competitors*. When

$$a_i > \frac{c_i}{2}, \frac{\partial p_i}{\partial h_i} > 0,$$

the price and expected share are a positive correlation. When

$$a_i < \frac{c_i}{2}, \frac{\partial p_i}{\partial h_i} < 0,$$

the price and expected share are a negative correlation.

(2) *Cross network externalities coefficient*. When

$$c_i^d > h_i, \frac{\partial p_i}{\partial a_i} > 0,$$

the price and cross network externalities coefficient are a positive correlation. When

$$c_i^d < h_i, \frac{\partial p_i}{\partial a_i} < 0,$$

the price and cross network externalities coefficient are a negative correlation.

These results further prove the existence and impact of critical mass in the platforms network. When

$$a_i = \frac{c_i}{2},$$

the platforms network reaches a critical mass.

2.2 Dynamic game model of price expectations

This article analyzes the competitive behavior of internet platforms from the angle of expected quantity games strategy previously. However, it is difficult to observe user scale of competitors in reality when platforms make specific decisions. Therefore, people concern more about competitors' recent price strategy. Then, the paper further analyzes competitive behavior of internet platforms from the angle of expected price games strategy. According to adaptive expectations in economics (time nearly effects of business or personal decisions), we amend competition model of platforms with joining the recent price adjustment factor.

Assuming price adjustment factor of platforms is influenced by the price of competitors in the previous period. The pricing function of platforms is

$$\begin{cases} p_{it}^d(n_{it}^s, p_{j(t-)}^d) = a_i^d n_{it}^s + h_i^d p_{j(t-1)}^d \\ p_{it}^s(n_{it}^d, p_{j(t-)}^s) = a_i^s n_{it}^d + h_i^s p_{j(t-1)}^s \end{cases}, \quad \begin{cases} p_{jt}^d(n_{jt}^s, p_{i(t-)}^d) = a_j^d n_{jt}^s + h_j^d p_{i(t-1)}^d \\ p_{jt}^s(n_{jt}^d, p_{i(t-)}^s) = a_j^s n_{jt}^d + h_j^s p_{i(t-1)}^s \end{cases} \tag{7}$$

Profit function of platforms can be transformed into

$$\begin{cases} \pi_i = (a_i^d n_{it}^s + h_i^d p_{j(t-1)}^d - c_i^d) n_{it}^d + (a_i^s n_{it}^d + h_i^s p_{j(t-1)}^s - c_i^s) n_{it}^s - C_i \\ \pi_j = (a_j^d n_{jt}^s + h_j^d p_{i(t-1)}^d - c_j^d) n_{jt}^d + (a_j^s n_{jt}^d + h_j^s p_{i(t-1)}^s - c_j^s) n_{jt}^s - C_j \end{cases} \tag{8}$$

Proposition 2: In the game about expected price, the price of one side set by platforms for users is determined by cross network externalities and expected adjustment of competitors' price. Final equilibrium results are

$$\begin{cases} n_{it}^d = \dfrac{c_i^s - h_i^s p_{j(t-1)}^s}{a_i^d + a_i^s} \\ n_{it}^s = \dfrac{c_i^d - h_i^d p_{j(t-1)}^d}{a_i^d + a_i^s} \\ n_{jt}^d = \dfrac{c_j^s - h_j^s p_{i(t-1)}^s}{a_j^d + a_j^s} \\ n_{jt}^s = \dfrac{c_j^d - h_j^d p_{i(t-1)}^d}{a_j^d + a_j^s} \end{cases}; \quad \begin{cases} p_{it}^d = \dfrac{a_i^d c_i^d + a_i^s h_i^d p_{j(t-1)}^d}{a_i^d + a_i^s} \\ p_{it}^s = \dfrac{a_i^s c_i^s + a_i^d h_i^s p_{j(t-1)}^s}{a_i^d + a_i^s} \\ p_{jt}^d = \dfrac{a_j^d c_j^d + a_j^s h_j^d p_{j(t-1)}^d}{a_j^d + a_j^s} \\ p_{jt}^s = \dfrac{a_j^s c_j^s + a_j^d h_j^s p_{j(t-1)}^s}{a_j^d + a_j^s} \end{cases}. \tag{9}$$

From these results, the users' scale of one side on the platform is decided by the marginal cost of providing these service, prior price set by competitors and cross network externalities. The price of one side on the platform is depended on cross network externalities, the expected price of competitors and marginal cost.

(1) *Cross network externalities coefficient.* Assuming that the expected price of competitors and marginal cost are constant, the price function is

$$p_i^d = \frac{a_i^d \overline{c_i^d} + a_i^s h_i \overline{p_{j(t-1)}^d}}{a_i^d + a_i^s} \quad \text{(where } h_i > 0\text{)}.$$

When $(a_i^d \overline{c_i^d} + a_i^s h_i \overline{p_{j(t-1)}^d}) > (a_i^d + a_i^s)$, the larger cross network externalities coefficient, the lower the price set by platforms. When $(a_i^d \overline{c_i^d} + a_i^s h_i \overline{p_{j(t-1)}^d}) < (a_i^d + a_i^s)$, the larger the cross network externalities coefficient, the higher the price set by platforms.

(2) *Marginal cost.* Assuming that cross network externalities coefficient and the expected price of competitors are constant, the price function is

$$p_i^d = \frac{\overline{a_i^d}\,\overline{c_i^d} + \overline{a_i^s}\,\overline{h_i p_{j(t-1)}^d}}{\overline{a_i^d} + \overline{a_i^s}}$$

It means that the higher the cost of these services provided for each customer interaction, the higher the price charged by platforms.

(3) *Expected price of competitors*. Assuming that cross network externalities coefficient and marginal cost are constant, the price function is

$$p_i^d = \frac{\overline{a_i^d}\,\overline{c_i^d} + \overline{a_i^s}\, h_i p_{j(t-1)}^d}{\overline{a_i^d} + \overline{a_i^s}} \quad \text{(where } h_i > 0\text{)}.$$

It means that the price set by platforms is a positive correlation with the expected price on the other side of competing platforms.

Also it is known that platforms can set a price on their service based on the size of externalities, and then enter the competitive differentiation stage, after critical capacity.

Corollary 2: In repeated games, the equilibrium price of internet platforms is

$$p_{it}^d = \frac{(a_i^d c_i^d + a_i^s h_i^d) a_j^d c_j^d + (a_j^d + a_j^s) a_i^d c_i^d}{(a_i^d + a_i^s)(a_j^d + a_j^s) - (a_i^d c_i^d + a_i^s h_i^d) a_j^s h_j^d}. \tag{10}$$

This article analyzes the competitive strategies between the two platforms. To simplify the analysis, the interaction coefficient of the same platform does not distinguish, namely $a_i^d = a_i^s$, $a_j^d = a_j^s$. Thus, the formula can be simplified to:

$$p_i = \frac{2(c_i + h_i) a_i a_j c_j + 2a_j a_i c_i}{4a_i a_j - 2(c_i + h_i) a_i a_j h_j} = \frac{(c_i + h_i) c_j + c_i}{2 - (c_i + h_i) h_j}. \tag{11}$$

The price of the platforms is affected by the marginal cost and expected adjustment factor.

According to adaptive expectations, the expected price is mainly affected by the previous one, that is $h_i = h_j = 1$, the price function can be translated into

$$p_i = \frac{(c_i + 1) c_j + c_i}{2\text{-}(c_i + 1)*1} = \frac{c_i c_j + c_j + c_i}{1\text{-}c_i} \tag{12}$$

Partial derivative:

$$\begin{cases} \dfrac{\partial p_i}{\partial c_i} = \dfrac{1}{(1\text{-}c_i)^2} \\ \dfrac{\partial p_i}{\partial c_j} = \dfrac{1 + c_i}{1\text{-}c_i} \end{cases}. \tag{13}$$

As

$$\frac{\partial p_i}{\partial c_i} > 0,$$

the price of a platform is a positive correlation with its marginal cost, namely the higher the marginal cost of the platforms, the higher the tariffs will be charged. As for the effect of others' cost, when $c_i > 1$ the price of a platform is a negative correlation with marginal cost of competitors; when $c_i < 1$ the price of a platform is a positive correlation with marginal cost of competitors. This means platforms have the character of critical capacity.

3 Conclusion

According to our analysis, internet platforms are mainly based on the size of the network externalities to conduct game competition [14]. There is a critical mass in the development and competition of platforms [15]. Before the critical mass, the role of external effects of the platform is very small, so it pursues rapid expansion of the size of the users, which can survive and avoid the establishment of failure. After the critical mass, the role of external effects of the platform is larger, so it adopts difference competition based on the size and the expected external behavior of competitors, which is in order to maximize corporate profits.

Therefore, internet platforms need to compete based on the stage.

First, before the critical mass, platforms can adopt a single product or services strategy in order to control costs, as much as possible to attract more customers. It uses cost minimization rather than differentiation strategy to pursuit rapid growth in user scale.

Second, after reaching critical mass, platforms can adopt a differentiation strategy based on network externalities size. It can better meet customers' different demands by providing different products or services for different customer groups, according to interactive coefficient.

Acknowledgements

The author is highly grateful to any anonymous referee for their careful reading and insightful comments, and the views and opinions expressed are those of the authors. The work is supported by the National Social Science Foundation of China (No. 14CGL012) and National Science and Technology Support Program (No. 2012BAH69F03).

References

[1] Rochet, J.C. & Tirole, J., Cooperation among competitors: some economics of payment card associations. *RAND Journal of Economics*, **33(4)**, pp. 549–570, 2002.

[2] Jullien, B.C.A.B., Chicken & egg: competition among intermediation service providers. *RAND Journal of Economics*, **34(2)**, pp. 309–328, 2003.
[3] Hagiu, A., Pricing and commitment by two-sided platforms. *RAND Journal of Economics*, **37(3)**, pp. 720–737, 2006.
[4] Shapiro, C. & Varian, H.R., Information rules: a strategic guide to the network economy. *Harvard Business Review*, pp. 152–198, 1999, ISBN: 7-300-03493-4/F.1048
[5] Katz, M.L. & Shapiro, C., Network externalities, competition, and compatibility. *American Economic Review*, **3(75)**, pp. 424–440, 1985.
[6] Rochet, J. & Tirole, J., Platform competition in two-sided markets. *Journal of the European Economic Association*, **1(4)**, pp. 990–1029, 2003.
[7] Armstrong, M., Competition in two-sided markets. *RAND Journal of Economics*, **37(3)**, pp. 668–691, 2006.
[8] Economides, N., The economics of networks. *International Journal of Industrial Organization*, **14(2)**, pp. 1–33, 1996.
[9] Rochet, J. & Tirole, J., Two-sided markets: a progress report. *RAND Journal of Economics*, **37(3)**, pp. 645–667, 2006.
[10] Weyl, E.G., A price theory of multi-sided platforms. *American Economic Review*, **100**, pp. 1642–1742, 2010.
[11] Weyl, E.G. & White, A., Insulated platform competition. Working Paper, 2013, available at SSRN: http://dx.doi.org/10.2139/ssrn.1694317.
[12] Wright, J., The determinants of optimal interchange fees in payment systems. *Journal of Industrial Economics*, **52**, pp. 1–26, 2004.
[13] Fudenberg, D. & Peysakhovich, A., Recency, records and recaps: learning and non-equilibrium behavior in a simple decision problem, Working Paper, 2014, available at SSRN: http://dx.doi.org/10.2139 /ssrn.2339862.
[14] Rysman, M., The economics of two-sided markets. *Journal of Economic Perspectives*, **23(3)**, pp. 125–143, 2009.
[15] Evans, D.S., Platform economics: essays on multi-sided businesses. *Competition Policy International*, 2011.

A novel dwelling time design method for low probability of intercept in a complex radar network

Z. Zhang[1,2,3], J. Zhu[1] & S. Salous[3]
[1]Science and Technology on Electronic Information, Control Laboratory, People's Republic of China
[2]College of Electronic Information, JiangSu University of Science and Technology, People's Republic of China
[3]School of Engineering and Computing Sciences, Durham University, UK

Abstract

To achieve the important tactical requirement of low probability of intercept (LPI) in the complex radar network, dynamically controlling the emission of the radars is very necessary. A novel radar dwelling time control strategy based on an interacting multiple model algorithm is presented in this paper, which controls the dwelling time of radar according to predicted covariance matrix during tracking, taking advantage of the relation model between the dwelling time and the tracking performance. First, the complex radar network is built for target tracking. Secondly, the influence of the dwelling time is considered in the tracking performance of the complex radar network. Finally, a decision will be made after the dwelling time for every radar is obtained by particle swarm optimization, the radar with the smallest dwelling time will be selected to track target. The tracking accuracy and LPI performance are demonstrated in the Monte Carlo simulations. The results are validated through the comparison with other methods.
Keywords: dwelling time, interacting multiple model, low probability of intercept, target tracking.

1 Introduction

As we know, the less the emitted time of the radar, the more the excellent performance of the low probability of intercept (LPI) [1]. To achieve the important

www.witpress.com, ISSN 1755-8336 (on-line)
doi:10.2495/978-1-78466-155-7/012

requirement of LPI, dynamically scheduling and controlling the dwelling time of the radar during the sensor management is very necessary. The work in [2] formalizes the typical workload of a modern-phased array radar and proposes a rate-based approach to schedule radar dwells in a real-time manner. The paper [3] develops a generalized framework for the radar task scheduling problem as an optimization model, and all radar task parameters are treated as variables, thereby allowing greater scheduling flexibility and the ability to handle more targets using a single radar. The scheduling of track dwells to minimize radar energy and time with an agile beam radar is considered in [4], where the trade between higher energy waveforms and radar time is further investigated. The algorithm in [5] introduces time windows that specify allowable earliness and lateness of radar tasks, and proposes a chaining process that combines the dwell times and the time windows of tasks with consecutive priorities. The work of Severson and Paley [6] presents a distributed, consensus-based approach to optimize radar resource management for ballistic missile surveillance and tracking.

Almost all of those works concern the performance of the single radar, and there are few studies on the design of dwelling time and radars selection in the complex radar network for excellent LPI performance. In this paper, a novel scheduling algorithm for dwelling time in radar networks is proposed. The remainder of this paper is organized as follows. Sections 2 and 3 describe the interacting multiple model (IMM) algorithm of target tracking and particle swarm optimization (PSO), respectively. Section 4 presents the dwelling time scheduling method in sensor network in detail. Simulations of the proposed algorithms and comparison results with other methods are provided in Section 5. The conclusions are presented in Section 6.

2 Target tracking algorithm based on IMM

IMM method is used for tracking maneuvering target in this paper. Many dynamic models are used for matching different motion states, and the switch probability of a different model is a Markov chain. Kalman filtering is employed for the estimation of target state and the update of model probability in the algorithm, and all dynamic models are parallel processed and the model probability represents model switch [7].

All the dynamic models are: $M=\{m^1,m^2,...m^r\}$, m_k^i is the ith model used at time k, the switch probability from model m_k^j to model m_{k+1}^i is $P\{m_{k+1}^j \mid m_k^i\}=\pi_{ij}$, m_{k+1}^j, $m_k^i \in M$, $\sum_{i=1}^{r}\pi_{ij}=1, i=1,2,...,r$. $\mu_j(k)$ is the probability of model j, $\mu_j(k)=P\{m(k)=j \mid Z^k\}$.

Let $\boldsymbol{X}(k)$ and $\boldsymbol{Z}(k)$ represent the state vector and the observation vector respectively, the state equation and transfer equation at time k are:

$$\boldsymbol{X}(k+1)=\phi_j(k+1)\boldsymbol{X}(k)+\boldsymbol{w}_j(k) \tag{1}$$

and

$$Z(k) = H_j X(k) + v_j(k) \tag{2}$$

where $w_j(k)$ and $v_j(k)$ are stationary white-noise processes with covariance matrices $Q_j(k)$ and $W(k)$ of model j, $\phi_j(k)$ is the transition matrix and H_j is the observation matrix.

Every recurrence of the IMM algorithm contains: interacting of input, model's filtering, update of model probability and interacting of output.

2.1 Interacting of input

Utilizing all the states and model probabilities from last recurrence, the computation of input state $\hat{X}_{0j}(t_{k-1})$ and covariance $P_{0j}(t_{k-1})$ of model j can be expressed as:

$$\hat{X}_{0j}(k-1) = \sum_{i=1}^{r} \hat{X}_i(k-1)\mu_{i|j}(k-1) \tag{3}$$

$$P_{0j}(k-1) = \sum_{i=1}^{r} (P_i(k-1) + \alpha.\alpha^T).\mu_{i|j}(k-1) \tag{4}$$

$$\alpha = [\hat{X}_i(k-1) - \hat{X}_{0j}(k-1)] \tag{5}$$

$$\mu_{i|j}(k-1) = P\{M(k-1) = i \mid M(k) = j, Z^{k-1}\} = \frac{\pi_{ij}\mu_i(k-1)}{\sum_{i=1}^{r} \pi_{ij}\mu_i(k-1)} \tag{6}$$

2.2 Model's filtering

In the light of step (1) and observation data $Z(k)$, the Kalman filtering of every model is as follows:

The prediction of state and covariance for model j has the form of

$$\hat{X}_j'(k) = \phi_j(k)\hat{X}_{0j}(k-1) \tag{7}$$

$$P_j'(k) = \phi_j(k)P_{0j}(k-1)(\phi_j(k))^T + Q_j \tag{8}$$

The residual error represented by observation data and estimation of last observation is:

$$v_j(k) = Z(k) - H_j\hat{X}_j'(k) \tag{9}$$

The variance matrix of the residual error is obtained by the follow equation:

$$S_j(k) = H_j P_j'(k)(H_j)^T + W_j(k) \tag{10}$$

Computation of the filtering gain is given as:

$$K_j(k) = P_j'(k)(H_j)^T(S_j(k))^{-1} \tag{11}$$

Equation of updated state is:

$$\hat{\boldsymbol{X}}_j(k) = \hat{\boldsymbol{X}}'_j(k) + \boldsymbol{K}_j(k)\boldsymbol{v}_j(k) \tag{12}$$

Estimation of the covariance is:

$$\boldsymbol{P}_j(k) = \left(\boldsymbol{I}\text{-}\boldsymbol{K}_j(k)\boldsymbol{H}_j\right)\boldsymbol{P}'_j(k) \tag{13}$$

2.3 Updating of the model probability

The model probability is recursively updated as:

$$\boldsymbol{\mu}_j(k) = \frac{\Lambda_j(k)\sum_{i=1}^{r}\pi_{ij}\boldsymbol{\mu}_i(k-1)}{\sum_{j=1}^{r}\sum_{i=1}^{r}\Lambda_j(k)\pi_{ij}\boldsymbol{\mu}_i(k-1)} \tag{14}$$

$\Lambda_j(k)$ is the likelihood function of model j at time k.

2.4 Interacting of output

The final estimation of state vector from all the models' state is:

$$\hat{X}(k) = \sum_{j=1}^{r}\hat{\boldsymbol{X}}_j(k)\boldsymbol{\mu}_j(k) \tag{15}$$

Estimated matrix of covariance is derived:

$$\boldsymbol{P}(k) = \sum_{j=1}^{r}\boldsymbol{\mu}_j(k)(\boldsymbol{P}_j(k) + \boldsymbol{b}\boldsymbol{b}^T) \tag{16}$$

where $\boldsymbol{b} = [\hat{\boldsymbol{X}}_j(k) - \hat{\boldsymbol{X}}(k)]$.

3 Particle swarm optimization (PSO)

PSO algorithm proposed by Kennedy and Eberhart [8] has been widely used in many fields. It is an evolutionary algorithm according to social interaction between independent particles. In PSO method, every potential solution for optimization problem is supposed as a point called particle in the space. The swarm composed by N particles moves through the problem space with the moving velocity of each particle. The position and velocity of the ith particle is represented as $\boldsymbol{x}_i = (x_{i1}, x_{i2}, \dots, x_{iN})$ and $\boldsymbol{v}_i = (v_{i1}, v_{i2}, \dots, v_{iN})$, respectively, where N is the dimension of the space.

The ith particle tracks its personal best position (*pbest*) as a vector $\boldsymbol{p}_i = (p_{i1}, p_{i2}, \dots, p_{iN})$; the global best position (*gbest*) among all the particles is represented as $\boldsymbol{p}_g = (p_{g1}, p_{g2}, \dots, p_{gN})$.

Usually the optimization problem can be formulated as follows:

$$\min f(X),\ \text{subject to } g(x) \geq 0 \tag{17}$$

where $\boldsymbol{X}=(x_1, x_2, \ldots, x_m)$.

The steps of PSO method are outlined as follows:

1. Generating the original particle position and velocity.
2. Exceeding boundary control.
3. Computing the fitness function $f(X)$ for each particle.
4. Updating the $pbest_i$:

$$pbest_i^{k+1} = \begin{cases} x_i^{k+1} & if\ f(x_i^{k+1}) < f(pbest_i^k) \\ pbest_i^k & otherwise \end{cases} \tag{18}$$

5. Updating the $gbest_i$:

$$gbest_i^{k+1} = \begin{cases} pbest_i^{k+1} & if\ f(x_i^{k+1}) < f(pbest_i^k) \\ gbest_i^k & otherwise \end{cases} \tag{19}$$

6. Updating the position and velocity:

$$\begin{aligned} v_{id}^{k+1} &= w \cdot v_{id}^k + c_1 \cdot r_1 \cdot (pbest_{id}^k - x_{id}^k) \\ &+ c_2 \cdot r_2 \cdot (gbest_{id}^k - x_{id}^k) \end{aligned} \tag{20}$$

$$x_{id}^{k+1} = x_{id}^k + v_{id}^{k+1} \tag{21}$$

where $i = 1,2,\ldots n$, $d = 1,2\ldots m$, n and m are the number of particles and dimensions in the space, w represents the inertia weight factor which is between 0.4 and 0.9, c_1 and c_2 are acceleration constants usually selected as 2, r_1 and r_2 are uniformly distributed random numbers in (0, 1).

7. Return to step (2) and go on the iteration.

4 Dwelling time allocation algorithm based on PSO

4.1 Design for covariance matrix W_k of measurement noise

The covariance matrix W_k of measurement noise is controlled by the emitted energy. As we know, radar equation at time k is as follows:

$$R_k^4 = t_B^k \frac{P_{av}^k G_T G_R \lambda^2 \sigma_k}{(4\pi)^3 K T_R S_{NR}^k L} \tag{22}$$

where t_B^k is the single dwelling time of the beam from the normal direction at time k, P_{av}^k is the average radiated power, G_R is the receiver gain, σ_k is the radar cross section (RCS) of the target, K is Boltzmann constant, T_R and L are respectively effective noise temperature and radar system loss, R_k is the detection range, G_T is the transmit gain, S_{NR}^k represents the signal-to-noise ratio of the system at time k.

Suppose when the target whose range is R_0, the radar has to emit the power P_{av0}, and the radar equation becomes

$$R_0^4 = t_{B0} \frac{P_{av0} G_T G_R \lambda^2 \sigma_0}{(4\pi)^3 K T_R S_{NR0} L} \tag{23}$$

Combining (22) with (23), the emitted signal-to-noise ratio at time k can be written as

$$S_{NR}^k = \frac{t_B^k P_{av}^k S_{NR0}}{t_{B0} P_{av0}} \frac{R_0^4}{R_k^4} \tag{24}$$

where the emitted power is supposed to be a constant parameter in this paper.

The single pulse signal is radiated by the radar, and the covariance of the measurement noise can be denoted as

$$\boldsymbol{W} = \begin{bmatrix} \dfrac{c^2 T_p}{8 S_{NR}^k} & 0 \\ 0 & \dfrac{3c^2}{w_c^2 T_p^{\,2} S_{NR}^k} \end{bmatrix} \tag{25}$$

where T_p is the pulse width, c is the wave velocity, and w_c is the carrier frequency. We can see that different can lead to different $\boldsymbol{W}$. However, during the tracking process, R_k is unknown before radar detection. Therefore, R_k in (24) is replaced by R_k^{pre}, which is predicted by R_{k-1} and v_{k-1}. R_k^{pre} is presented as

$$R_k^{pre} = R_{k-1} + v_{k-1} T \tag{26}$$

R_{k-1} and v_{k-1} are the target's range and velocity, which are estimated by the IMM tracking algorithm at time $k-1$, and T is the tracking interval.

4.2 Computation of predicted tracking covariance matrix

The prediction of covariance for model j at time k can be represented as

$$\boldsymbol{P}_j^{'}(k+1) = \phi_j(k+1) \boldsymbol{P}_{0j}(k) (\phi_j(k+1))^T + \boldsymbol{Q}_j \tag{27}$$

Then, the variance matrix $S_j(k+1)$ and filtering gain $K_j(k+1)$ can be written, respectively,

$$\boldsymbol{S}_j(k+1) = \boldsymbol{H}_j \boldsymbol{P}_j^{'}(k+1)(\boldsymbol{H}_j)^T + \boldsymbol{W}_j(k+1) \tag{28}$$

$$\boldsymbol{K}_j(k+1) = \boldsymbol{P}_j^{'}(k+1)(\boldsymbol{H}_j)^T (\boldsymbol{S}_j(k+1))^{-1} \tag{29}$$

The covariance estimation for every model can be represented as

$$\boldsymbol{P}_j^{pre}(k+1) = (\boldsymbol{I}\text{-}\boldsymbol{K}_j(k+1)\boldsymbol{H}_j)\boldsymbol{P}_j^{'}(k+1) \tag{30}$$

At last, the predicted covariance matrix is given as

$$\boldsymbol{P}^{pre}(k+1) = \sum_{j=1}^{r} \boldsymbol{\mu}_j(k)(\boldsymbol{P}_j^{pre}(k)) \tag{31}$$

where $\mu_j(k)$ is model probability at time k.

4.3 Dwelling time allocation algorithm based on PSO

The desired tracking covariance matrix $\boldsymbol{P}_{des}$ should be set for the m radars in the network first. Then, the dwelling time of every radar is selected as the particles in the PSO, and then the minimum and maximum ranges, velocities are designed according to the radars' performance. Initial locations and velocities of all particles are generated randomly in whole search space.

Then, the fitness function is designed as

$$f = \mathbf{min}(\text{trace}(\boldsymbol{P}^{pre}(k+1) - \text{trace}(\boldsymbol{P}_{des}))) \tag{32}$$

During the PSO optimization, every radar in the network will select the dwelling time $T_{dw}{}^{m}$, which leads to the minimum *fitness function*. Then, a dwelling time set can be given as

$$\boldsymbol{T}_{dw} = \left\{\boldsymbol{T}_{dw}{}^{1}, \boldsymbol{T}_{dw}{}^{2} \ldots, \boldsymbol{T}_{dw}{}^{m}\right\} \tag{33}$$

To save most radiation time and get the best LPI ability of the radar network, the algorithm will choose the radar which will radiate the minimum dwelling time in the network for tracking at time k.

5 Simulation results

In this section, Monte Carlo simulations are performed to analyze the performance of the proposed dwelling time scheduling algorithm based on the PSO and predicted covariance (PSO-PC). The IMM filter here comprises constant velocity (CV) model and coordinated turn (CT) rate model [7].

5.1 Trajectory design

Figure 1 shows the target trajectory in 100s. There are three radars, which are labeled A, B and C in the radar network. The positions of the radars are (0.80), (250.10) and (-70,400), respectively.

5.2 Comparison of tracking performance

The proposed dwelling time design method (PSO-PC) of three radars is realized in the simulation, which is also compared with the performance of every single radar using the scheduling strategy of dwelling time in [6]. The root mean square error (RMSE) of time k and the average root-mean-square error (ARMSE) of the whole tracking process can be formulated as (34) and (35), respectively:

$$\boldsymbol{RMSE}(k) = \sqrt{\frac{1}{M_c}\sum_{m=1}^{M_c}(x_k - \hat{x}_k^m)^2} \tag{34}$$

$$ARMSE = \frac{1}{N_t}\sum_{k=1}^{N_t}\boldsymbol{RMSE}(k) \tag{35}$$

where M_c is the number of the Monte Carlo simulation, x_k is the true state of the system, is the estimated vector at the mth simulation, and N_t is the total tracking time in every simulation. In the simulation, M_c=100 and N_t=100.

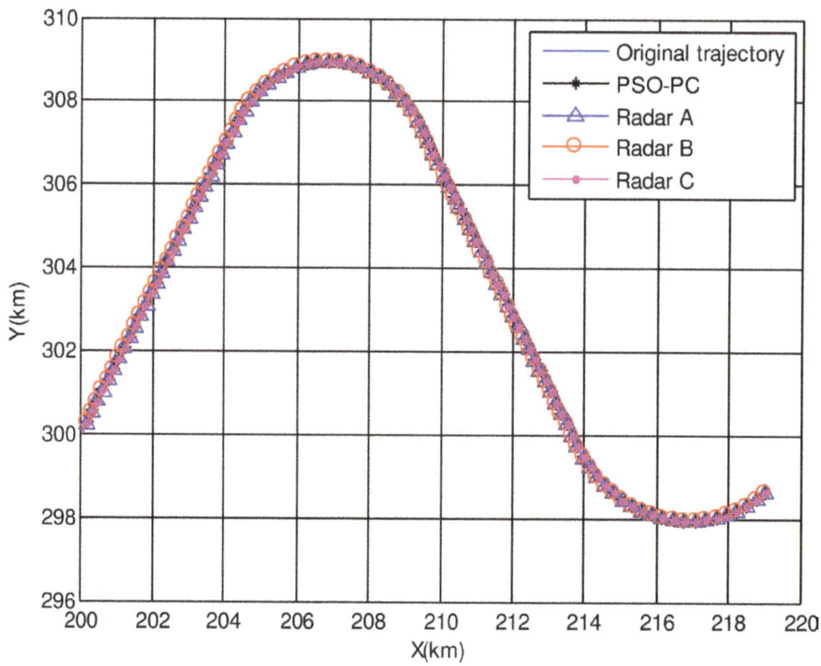

Figure 1: Trajectory of the target.

Figure 2 shows the range RMSE of the proposed method during the tracking. Figure 3(a) and (b) shows the RMSE of the proposed method in X and Y direction, respectively. Table 1 shows the ARMSE of the range, X direction, and Y direction. Compared with every radar, we can see that the proposed method of PSO-PC presents almost the same excellent tracking accuracy.

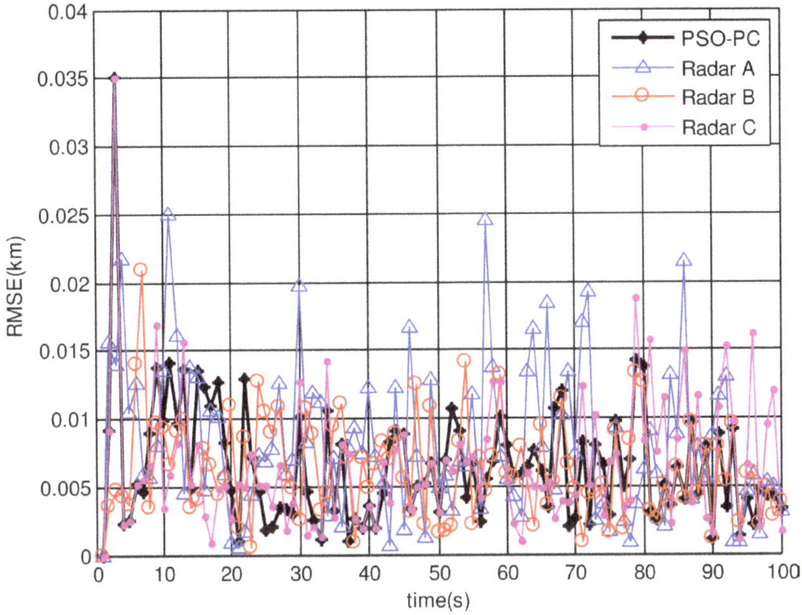

Figure 2: Comparison of tracking performance.

(a)

(b)

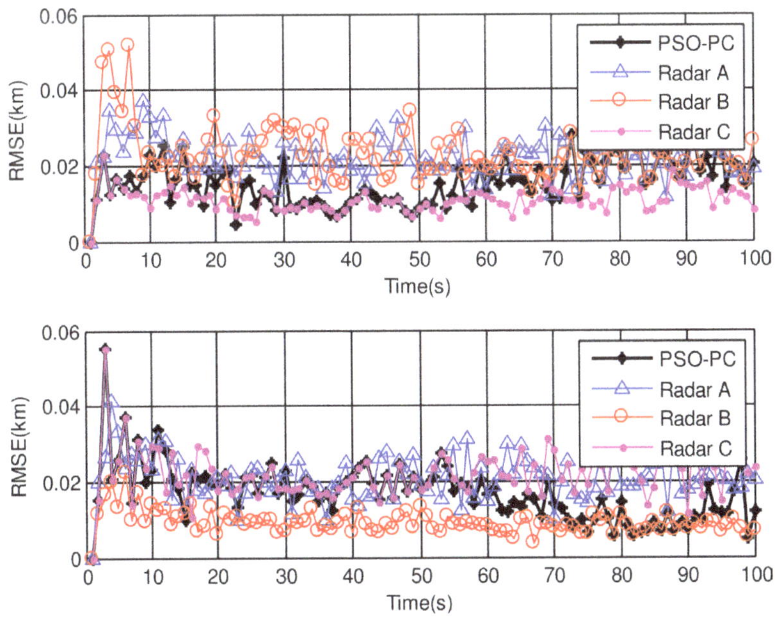

Figure 3: Comparison of tracking performance: (a) tracking performance of X direction and (b) tracking performance of Y direction.

Table 1: Comparison of ARMSE.

Method	R-ARMSE (km)	X-ARMSE (km)	Y-ARMSE (km)
PSO-PC	**0.0065**	**0.0155**	**0.0169**
Radar A	0.0083	0.0224	0.0210
Radar B	0.0063	0.0229	0.0091
Radar C	0.0066	0.0109	0.0223

5.3 Comparison of dwelling time

As dwelling time of PSO-PC, Radar A, Radar B and Radar C is shown in Fig. 4, we can see that the proposed method not only presents excellent tracking accuracy but also reduces more radiated time. As Fig. 5 shows the radiation label of the radars, we can see the radars work in turn in order to obtain excellent tracking and LPI performance.

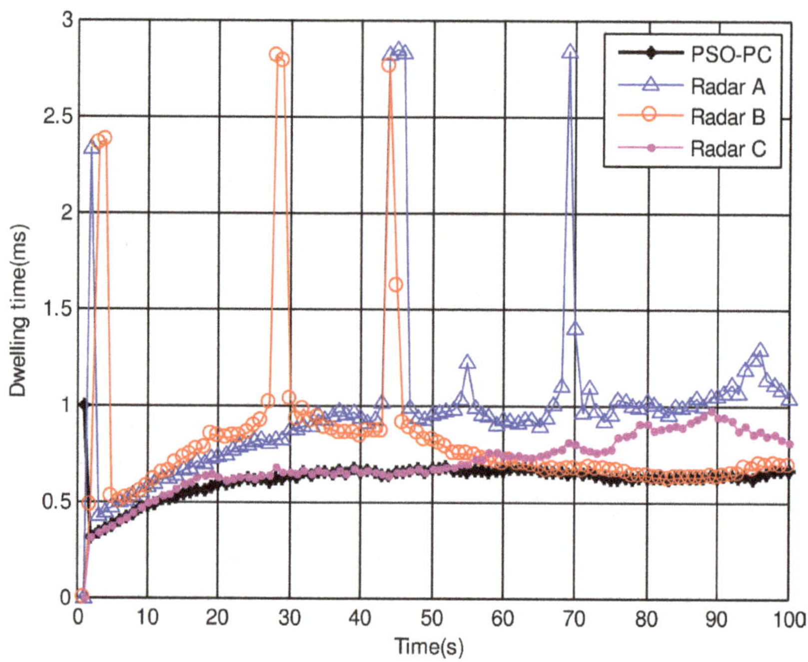

Figure 4: Comparison of dwelling time.

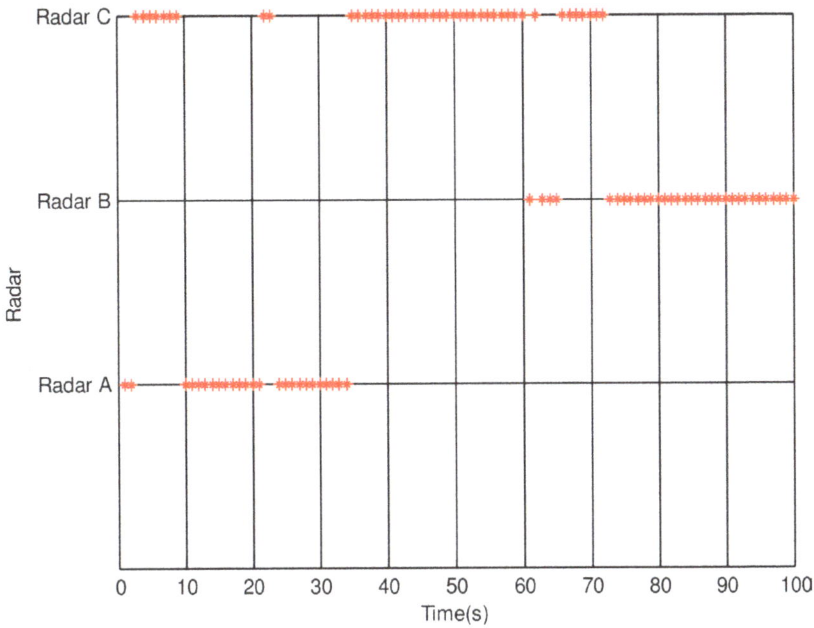

Figure 5: Radiation label of the radars.

6 Conclusions

In this paper, we have presented a new strategy of dwelling time allocation in the radar network based on the PSO method and predicted covariance theory. According to the optimization results of every radar, the radar with the minimum dwelling time in the network will be selected to track the target. The simulation results show that the proposed method can save much more dwelling time with excellent tracking accuracy for tracking single target in the network. As for the future work, this algorithm can be modified to track more targets in complex radar network.

Acknowledgements

This work was supported by the National Natural Science Funds Fund (61401179), Colleges of Jiangsu Province Natural Science Fund (14KJB510009), the Science and Technology on Electronic Information Control Laboratory Project, Scientific Research Start-up Funding from Jiangsu University of Science and Technology, Fundamental Research Funds for the Central Universities (NJ20140010), and a Project Funded by the Priority Academic Program Development of Jiangsu Higher Education Institutions.

References

[1] Krishnamurthy, V., Emission management for low probability intercept sensors in network centric warfare. *IEEE Transactions on Aerospace and Electronics Systems*, **41(1)**, pp. 133–151, 2005.

[2] Kuo, T.-W., Chao, Y.-S., Kuo, C.-F. & Chang, C., Real-time dwell scheduling of component-oriented phased array radars. *IEEE Transactions on Computers*, **54(1)**, pp. 47–60, 2005.

[3] Mir, H.S. & Guitouni, A., Variable dwell time task scheduling for multifunction radar. *IEEE Transactions on Automation Science and Engineering*, **11(2)**, pp. 463–472, 2014.

[4] Glass, J.D., Blair, W.D. & Bar-Shalom, Y., Optimizing radar signal to noise ratio for tracking manoeuvring targets. *2014 17th International Conference on Information Fusion (FUSION)*, Salamanca, 7–10 July, 2014, pp. 1–7.

[5] Jang, D.-S., Choi, H.-L. & Roh, J.-E., A time-window-based task scheduling approach for multi-function phased array radars. *2011 11th International Conference on Control, Automation and Systems (ICCAS)*, Gyeonggi-do, 26–29 October, 2011, pp. 1250–1255.

[6] Severson, T.A. & Paley, D.A., Distributed multitarget search and track assignment with consensus-based coordination. *IEEE Sensors Journal*, **15(2)**, pp. 864–875, 2015.

[7] Zhang, Z., Zhou, J., Wang, F., Liu, W. & Yang, H., Multiple-target tracking with adaptive sampling intervals for phased-array radar. *Journal of Systems Engineering and Electronics*, **22(5)**, pp. 760–766, 2011.

[8] Kennedy, J. & Eberhart, R., Particle swarm optimization. *Proceedings of IEEE International Conference on Neural Networks*, April, 1995, pp. 1942–1948.

Systems analysis for energy systems using an integrated model of GIS and technology models

H. Hamasaki
Fujitsu Research Institute and Centre for International Public Policy Studies, Japan

Abstract

This paper focuses on large-scale renewable electricity production and aims to identify key factors which affect renewable energy popularization, Japan's carbon emissions, energy independence, and system costs. Based on these factors, we will propose a new energy system for an alternative energy era, using a detailed sub-regional electricity technology model, the Japan Multi-regional Transmission (JMRT) model. Intermittency and geographical sensitivity are the two main factors that differentiate renewable electricity from conventional sources. Seasonal and diurnal variations in wind/solar electricity necessitate the use of backup capacity and storage. Furthermore, the most renewable source in Japan – onshore wind – has better potential in regions with low electricity demand. This makes integrating the more or less isolated (10) grids of Japan a very important issue. The JMRT model employs 1 km^2 grid GIS information on wind speeds, distances from the nearest road and from electricity grid for a very detailed description of wind potential.

This project uses a systems analysis approach, which is the dissection of a system into its component pieces to study how those pieces interact and work together.
Keywords: GIS, renewable energy, technology model.

1 Introduction

Japan currently faces many energy-related issues, of which energy independence and carbon emissions are two important policy targets. Most of the fossil fuels used in Japan are imported, and the country's energy self-sufficiency rate is a mere 4% (18% if nuclear power is included). The Fukushima disaster made the

WIT Transactions on State of the Art in Science and Engineering, Vol 90, © 2016 WIT Press
www.witpress.com, ISSN 1755-8336 (on-line)
doi:10.2495/978-1-78466-155-7/013

situation much worse by eliminating the nuclear option, which could have played a significant role in achieving both objectives for the country. Furthermore, the disaster has created a much more urgent and important objective, i.e. maintaining a reliable and economically viable energy supply for its people and industry, as across the nation nuclear plants are being shut down with immediate effect. Japan's electricity production mix is shown in Fig. 1. The 25% share of nuclear will be almost entirely met by fossil fuels – mainly LNG – over the short term. The current situation has worsened in terms of both self-sufficiency of energy and CO_2 emissions.

Fossil fuel thermal power stations with CCS (carbon capture and storage) and renewable energies can help in the climate objective, but only renewables would begin to move Japan towards self-sufficiency.

The objective of this paper is to characterize the renewable potential in Japan and to study its interactions with competitors under some key policy and technology scenarios.

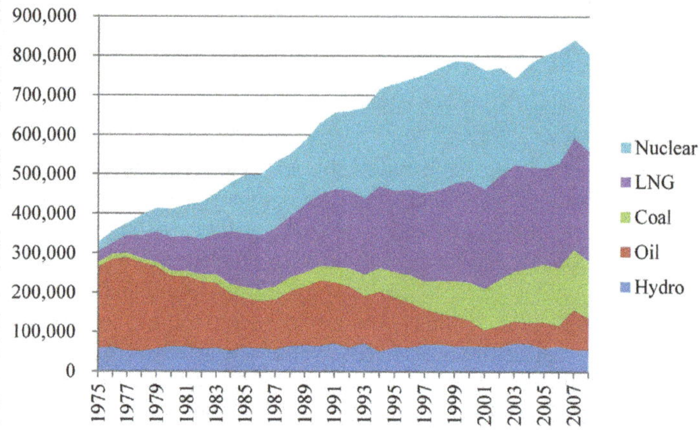

Figure 1: Japanese electricity mix (TWh). Source: The Institute of Energy Economics (2011), Handbook of Energy & Economic Statistics in Japan.

2 Bottom-up model (Japan Multi-Regional Transmission (JMRT) model)

The bottom-up model used for this research, the JMRT model, is a detailed disaggregate Japanese electricity generation system model based on TIMES to reflect factors which affect the popularization of renewables. The TIMES objective is to minimize the total cost of the system, properly augmented by the 'cost' of lost demand. All cost elements are appropriately discounted to a selected year [1].

2.1 Model structure

Japan has 10 electricity grids with weak connections between grids as shown in Fig. 2. In addition, there are two different electricity frequencies, 50 and 60 Hz, with frequency converters to convert one frequency to another as shown.

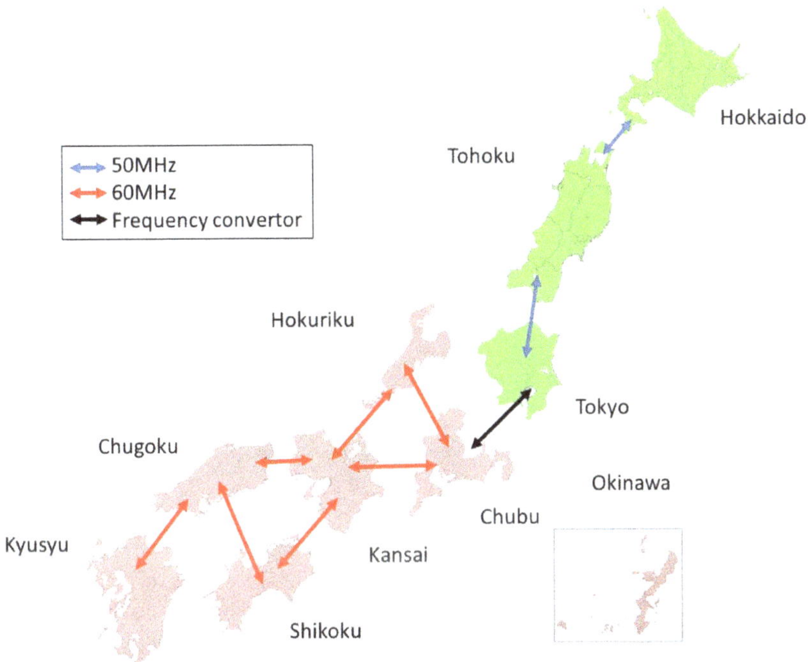

Figure 2: Grid system in Japan.

Figure 3 represents the energy system used in the model. The model focuses mainly on electricity supply, with existing power stations and pumped storage data included therein. The model assumes conventional power stations (Ultra-super Critical, Integrated Gasification Combined Combustion, Gas Turbine Combined Cycle, and nuclear) and renewables (biomass, on-shore and off-shore wind turbine, photovoltaic (PV), geothermal and small hydro) as new technologies. We use cost data of conventional power station of Committee of Electricity Generation Cost Verification [2] and IEA/NEA [2, 3].

2.2 Database JMRT model

To build a renewable potential database for the JMRT model, we use MoE potential data [4] and make the following modifications.

In the case of offshore wind, we add GIS calculated data, distance from the nearest infrastructure and sea depth, and also new data sets calculated using GIS data and GIS data for each mesh (Fig. 4).

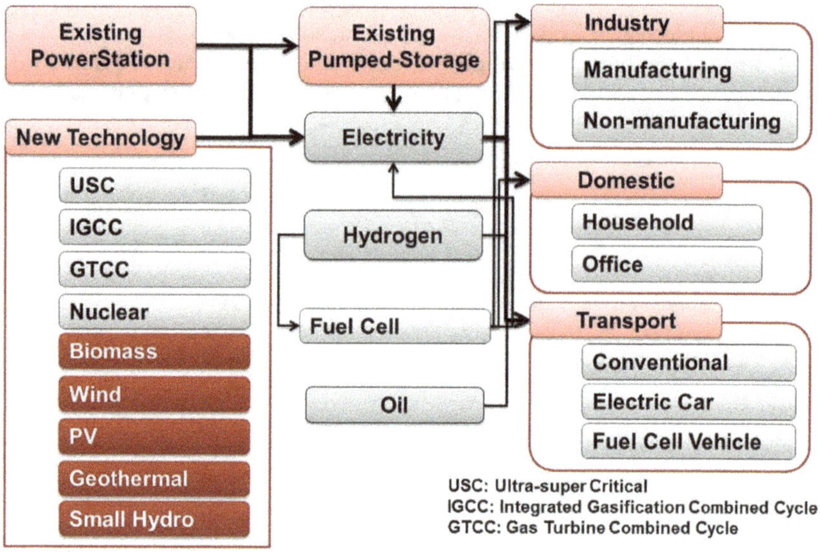

Figure 3: Energy system in JMRT.

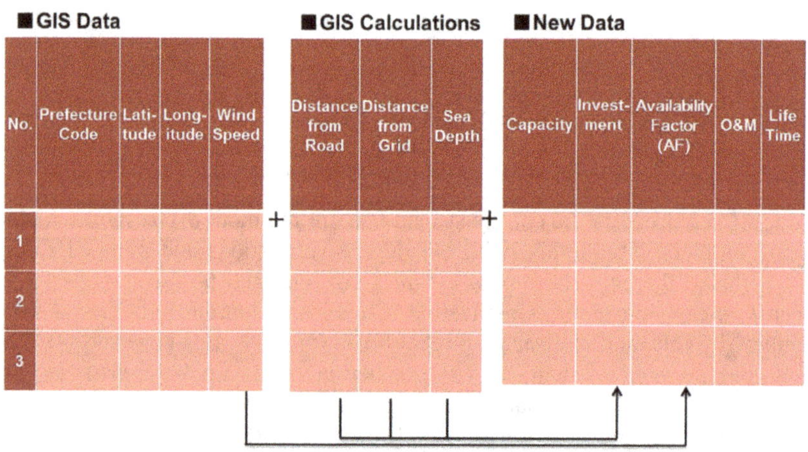

Figure 4: MOE GIS data and new data (case of off-shore wind).

For TIMES, we make clusters categorized by investment cost and availability factor, and the same clusters are applied to each prefecture. The upper limit of capacity installed in each cluster is applied based on the GIS data set as shown in Fig. 5.

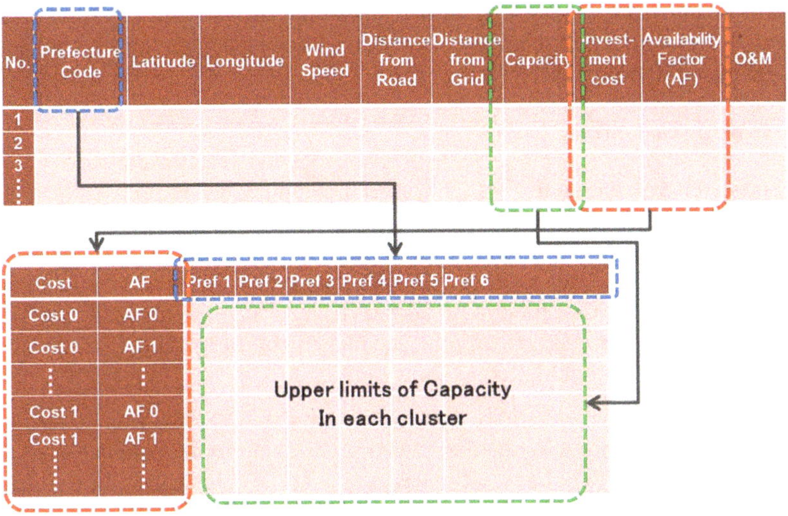

Figure 5: Conversion GIS data to TIMES.

3 Simulation

3.1 Simulation scenarios

In this paper, we employ systems analysis, a powerful tool for identifying the interactions between complex factors. We take two important policy objectives

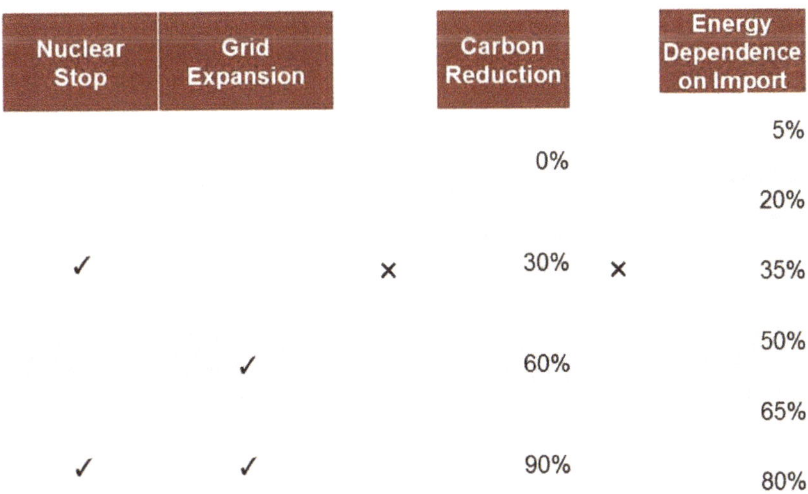

Figure 6: Design of systems analysis.

as factors, i.e. nuclear stop and grid expansion, and we multiply them with different levels of carbon reduction and energy dependence on import. The total number of simulations is 96.

3.2 Realistic cost curve

The supply curves shown in Fig. 7 have been computed using used 1 km^2 mesh GIS data for onshore and offshore wind, and prefectural data for solar PV. More than 1,000 TWh seem be available for well under 15 cents/kWh based on this figure. However, these costs and availabilities ignore very important spatial and temporal issues, which we propose to address using systems analysis.

There are two main reasons why the actual availabilities would turn out lower and prices would turn out higher, as shown in Fig. 8.

- The Japanese electricity system comprised 10 grids with very poor interconnections. While the costs of connecting to the regional grids are included in the above estimates, the costs of increasing inter-regional connections are not. Renewable potential is not sufficiently high on the grids in regions that have high demand. Therefore, using the lower part of the above curves necessitates grid expansion.
- The above prices assume full utilization of power when it is available. However, due to seasonal and diurnal variations in demand, this is not possible, and the actual utilization turns out lower than the potential.

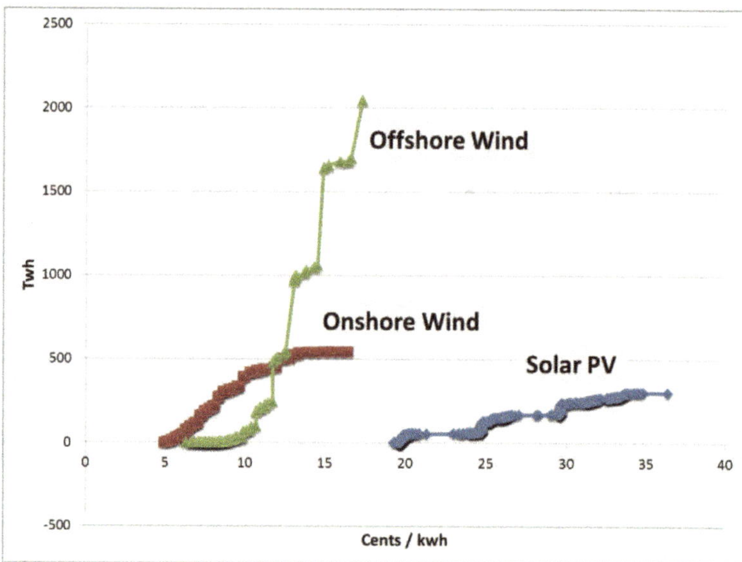

Figure 7: Supply curves of renewable energies.

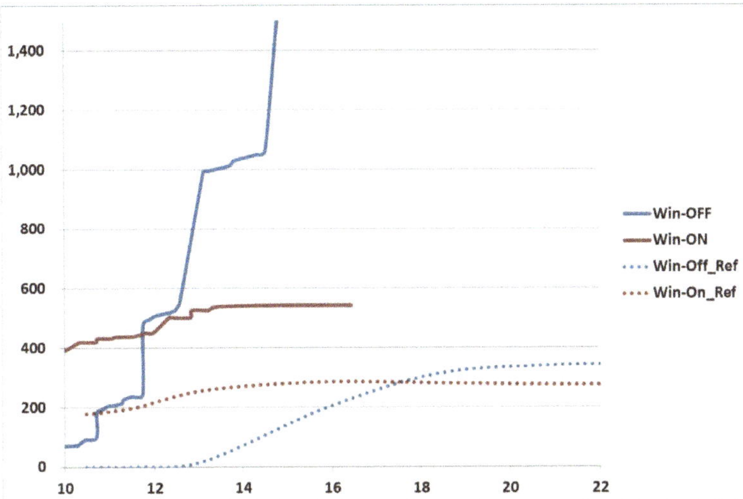

Figure 8: Realistic cost curves under current Japanese electricity system. Note: Win-OFF: potential of offshore wind, Win-ON: potential of onshore wind, Win-Off_Ref: reference of offshore wind, Win-On_Ref: reference of onshore wind.

Figure 9 shows energy and wind potential in each grid. Huge wind potential is evident in Hokkaido and Kyushu, but significant electricity consumption can be seen in Tokyo: in a word, there is a sizeable geological gap between wind potential and electricity consumption. Due to the limited connections between regions, it is impossible to make the most use of huge onshore wind potential in Hokkaido.

Grid expansion would make it possible to access and build wind turbines at sites with good wind conditions, and as a result, the actual availability would be even greater than the reference case, as shown in Fig. 10.

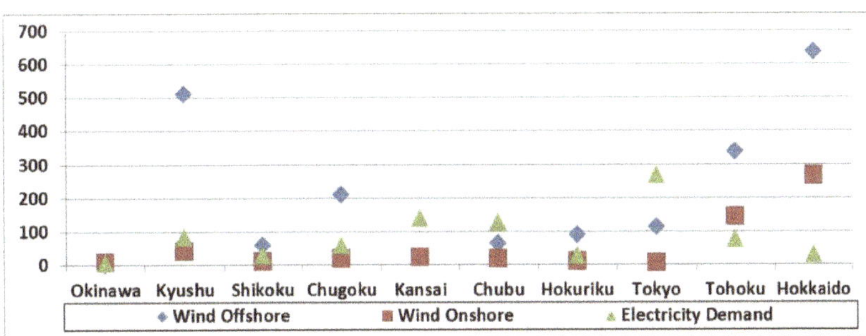

Figure 9: Energy consumption and wind potential in each grid.

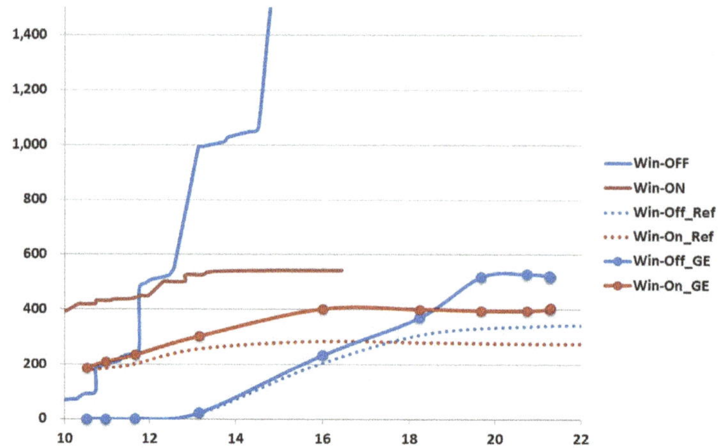

Figure 10: Comparison of cost curves of potential, actual, and grid expansion. Note: Win-OFF: potential of offshore wind, Win-ON: potential of onshore wind, Win-Off_Ref: reference of offshore wind, Win-On_Ref: reference of onshore wind, Win-Off_GE: grid expansion case of offshore wind, Win-On_GE: grid expansion case of onshore wind.

4 Conclusions

In this study, we have used a very detailed representation of Japan's current electricity system, along with very rich data on renewable potential, to explore the interplay between various electricity options for Japan. This work could be useful for policy makers as well as modellers. Modellers, who operate at a higher level of disaggregation than what we have shown in this paper, could use the deployment patterns exposed here as a reference. Quantitative estimates of the impacts of various policy options on primary objectives could be useful for policy makers.

As has been explained above, Japan is now facing a challenge which no other country has faced before: to raise its energy self-sufficiency ratio and reduce GHG simultaneously, even though most of its fossil fuel supply comes from overseas and it cannot depend upon nuclear power. Though these issues cannot be solved by existing technology, our analysis shows what needs to be done in the short term using existing technology and what direction R&D for new technology should pursue over the long term. In the near future, Japan should integrate all of its electricity transmission lines and create a single grid. Doing so would enable the country to maximize the use of cheaper renewable energy sources and increase the overall efficiency of their usage. CCS at coal-fired thermal power stations could be better utilized as well. A Feed-in Tariff system, which assures suppliers of renewable energy sources that the energy will be purchased at a fixed price, was introduced in 2012 in Japan to encourage utilization of renewable energy sources. However, with separated transmission networks, such a system could in fact raise the price of renewables and thus even be an impediment to their spread.

In the long run, even with a single grid it would be impossible to achieve a stable supply of wind and solar power as provided by Mother Nature. There would be a variety of views upon the extent to which renewable energy sources could be introduced because of their technological constraints. Japan would need extra facilities to meet demand at peak times in order to achieve a sufficient supply of electricity generated from renewables. In this research, we suggest the hydrogen cycle as an innovative concept, and the simulation results show that the hydrogen cycle could play a major role in achieving extreme carbon constraint and high energy independency by 2050.

References

[1] Loulou, R., Remne, U., Kanudia, A., Lehtila, A. & Goldstein, G., Documentation for the TIMES Model PART I. *Energy Technology Systems Analysis Programme*, 2005.
[2] Committee of Electricity Generation Cost Verification, National Policy Unit, Cabinet Secretariat, Japan, *Committee of Electricity Generation Cost Verification Report*, 19 December 2011.
[3] IEA/NEA, *Projected Costs of Generating Electricity*, 2010, IEA/NEA.
[4] Ministry of the Environment, *Renewable Energy Potential Survey Report*, 2011, Ministry of the Environment.

Analysing the Chinese stock market using the Hurst exponent, fractional Brownian motion and variants of a stochastic logistic differential equation

O. Vukovic
University of Liechtenstein, Liechtenstein

Abstract

The Chinese stock market is rapidly developing and is becoming one of the wealth management investment centres. Recent legislation has allowed wealth management products to be invested in the Chinese stock market. By taking data from St. Louis Fed and analysing the Chinese stock market using the Hurst exponent, which was calculated by using two methods, and fractional Brownian motion, it is proved that the Chinese stock market is not efficient. However, further analysis was directed to finding its equilibrium state by using logistic difference and a differential equation. To achieve more precise movement, a stochastic logistic and delayed logistic differential equation have been implemented which are driven by fractional Brownian motion. As there is no explicit solution to a delayed differential equation using the Stratonovich integral, a method of steps has been used. A solution has been obtained that gives the equilibrium state of the Chinese stock market. The following solution is only for a Hurst exponent that is higher than 1/2. If Hurst exponent is close to 1/2, then Brownian motion is an ordinary one and the equilibrium solution is different. Sensitivity analysis in that case has been conducted in order to analyse possible stationary states. The main conclusion is that the Chinese stock market is still not mature enough to be efficient, which represents a hidden risk in wealth management investing. A new approach to valuing the efficiency of the market by using nonlinear dynamics and system analysis has been proposed.
Keywords: Chinese stock market, fractional Brownian motion, Hurst exponent, stochastic logistic differential equation.

WIT Transactions on State of the Art in Science and Engineering, Vol 90, © 2016 WIT Press
www.witpress.com, ISSN 1755-8336 (on-line)
doi:10.2495/978-1-78466-155-7/014

1 Introduction

Wealth management products (WMP) are bank-generated investment products that are sold to their retail and institutional customers. Similar to money market funds, WMPs pool funds with a relatively short investment horizon (as short as 5 days) and invested them in longer duration assets to arbitrage the difference in returns. Disclosure on these products, including what assets they invested in and the likely returns of each, is limited. WMPs essentially circumvent China's tight control of interest rates by rewarding investors' higher returns than deposit rates. Recently, it has been allowed Chinese banks to invest money directly to stock and bond markets. It is the aforementioned decision that could have high potential consequences for the Chinese middle market as the stock and bond market is characterised by nonlinearities and, therefore, nonlinear dynamics can be applied to prove that Chinese stock market is characterised by nonlinear model that is more effective than the Bachelier's random walk theory based on stochastic models. The complex stochastic behaviour of a chaotic system is determined by the nonlinear equations of a determined system that has finite independent free variables, and the phase space is a finite dimension status space that is constructed by these independent variables. To analyse the system, the level of noise must be analysed. It is tried to subject stock price movement of Chinese market to nonlinear analysis. Hurst exponent is presented and its analysis is conducted to analyse nonlinearity level in the Chinese stock market. At the same time, there is a connection between Hurst exponent and fractional Brownian motion. If the movement of Chinese stock market can be explained by using fractional Brownian motion or Gaussian process, it can be concluded that market is not predictable and inefficient. Fractional Brownian motion is characterised by a non-semimartingale process and therefore cannot be integrated. However, if Hurst exponent has a value of 1/2, then Brownian motion is characterised as semimartingale and as it is possible to integrate with some type of stochastic integral, it is then possible to prove that the movement of Chinese stock market can be decomposed to a local martingale and cadlag-adapted process of bounded variation. As the definition adapted process is that it cannot be predicted, it is obvious that the value of Hurst exponent of 1/2 causes Chinese stock market to be unpredictable. Before going to real analysis and generated results, background concerning stochastic analysis as well as Hurst exponent will be presented. Main intention is to present the fact that investing in Chinese stock and bond markets can present risks for wealth management in China because if market is inefficient, it can be proved that according to ergodic theory, it is impossible to predict its state. Although according to Hairer [1], there are two theorems that state, if Hurst parameter has approximate value of 1/2, it can be predicted that the solution of stochastic differential equation converges to stationary value taking into consideration some initial conditions. As stochastic differential equations entail Brownian motion, therefore as fractal Brownian motion is being considered, it is possible that by using some form of stochastic differential equation, it can be proved that the movement of Chinese stock market converges to some specific value. To prove the following assumption, delayed logistic equations is used. It is known that ordinary logistic equation is being used to model population dynamics. In this paper, first logistic equation is being analysed. After that, Brownian motion is implemented in delayed logistic equation and approximate solution is presented. As stochastic analysis presents Ito integral and Stratonovich integral, it is presented that equation using Stratonovich integral cannot be solved. As ordinary logistic difference equation is characterised by cobweb diagram and attractor, it is tried to find a solution for delayed logistic equation and find its attractor and equilibrium value. Afterwards, delayed stochastic differential equation including Brownian motion is included. Its approximate solution is found and it is simulated to find the equilibrium point which is compared with the current state of Chinese stock market in order to see if it attains its equilibrium value. The results are discussed afterwards. In this manner, risk is analysed to see if wealth management has one of the global risks and that is investing in a market which movement cannot be predicted. Although the aforementioned approach is really unconventional, some results are really interesting and hopefully the provided results will be helpful to forward research in the following direction. Depending on the value of Hurst exponent, it can be concluded that if there is a high value of standard deviation and Hurst exponent value is 1/2, delayed stochastic logistic equation is used to model movement of stock price. It will be showed that because of high volatility, it is very hard to predict stock price movement on the Chinese stock market.

2 Background

Fractional Brownian motion is defined as continuous Gaussian process with $B_H(t)$ on $[0, T]$, which starts at 0, has expectation 0 for all t in $[0, T]$, and has the following covariance function [3]:

$$E[B_H(t)B_H(s)] = \tfrac{1}{2}(|t|^{2H} + |s|^{2H} - |t-s|^{2H}) \tag{1}$$

where Hurst exponent has the value between (0,1). The value of Hurst exponent determines what kind of movement will the time series exhibit. Hurst exponent can have the value between (0,1). If Hurst exponent is higher than 1/2, then time series exhibit positive correlation; if Hurst exponent has the value less than 1/2, then the series exhibits negative correlation, while for the value of Hurst exponent of 1/2, series exhibits ordinary Brownian motion. To introduce the stochastic differential equation, few assumptions will be made:

1. Stability;
2. Non-degeneracy;
3. Growth and regularity.

2.1 Stability

$$C^{(A_1)} > 0, \tag{2}$$

$$\langle f(x) - f(y), x - y \rangle \le \min\left\{ C_1^{(A_1)} - C_2^{(A_1)} \|x-y\|^2, C_3^{(A_1)} \|x-y\|^2 \right\} \tag{3}$$

for every $x, y \in R_n$ stability is the only condition used to limit the movement in a time series.

2.2 Growth and regularity

There exist constants $C, N > 0$ such that f and its derivatives satisfy the following condition:

$$\|f(x)\| \leq C\left((1+\|x\|)\right)^N \qquad \|Df(x)\| \leq C\left((1+\|x\|)\right)^N \tag{4}$$

For every $x \in R^n$, this condition only pertains to the limited growth of model that is controlled by the constant and parameter N. The derivative says that the growth of the function is limited and is given in one determined range.

2.3 Non-degeneracy

The $n \times n$ matrix of s is invertible.

As we will be considering the SDEs, it is recommendable to limit to the following equations of the form:

$$\mathrm{d}x_t = f(x_t)\mathrm{d}t + \sigma \mathrm{d}B_H(t) \tag{5}$$

where $x_t \in R^n$, $f : R^n \rightarrow R^n$, and B_H is an n-dimensional FBM with Hurst parameter that will be analysed. H and σ are constant and invertible $n \times n$. This should be interpreted as an integral equations. For the Hurst parameter of 1/2, the following equation becomes ordinary stochastic differential equation as Hurst parameter of 1/2 for fractional Brownian motion proves that Brownian motion is semimartingale and it behaves as random walk. FBM is self-similar with the scaling law

$$t \rightarrow B_H(at) \approx t \mapsto a^H B_H(t) \tag{6}$$

It appears that FBM can be used to model long-time correlations in stock markets. The reason for choosing fractional Brownian motion for modelling is that the value of Hurst exponent higher than 1/2 presents genuine long-term correlations that persist even under rescaling. Also, there exists simple formula that relates fractional Brownian motion to standard one, and there this is a useful result.

When Hurst exponent has higher value than 1/2, more regularity for f is required and it will be assumed that the stronger condition for growth and regularity holds.

The derivative of f is globally bounded.

Hairer in his paper on 'Ergodicity of stochastic differential equations driven by fractional Brownian motion' proves that the above-defined stochastic differential equations possess a unique stationary solution. Two theorems will be quoted, for detailed proof look at the Hairer aforementioned paper [1, 5, 6]:

T1 (Small Hurst parameter). Let $H \in (0, 1/2)$ and let f and σ satisfy the above-presented conditions. Then, for every initial condition, the solution to stochastic differential equations converges toward a unique stationary solution in the total variation norm. The difference between stationary solution is bounded by $C_\gamma t^{-\gamma}$ for large t for every $\gamma < \max_{\alpha<H} \alpha(1-2\alpha)$.

T2 (Large Hurst parameter). Let $H \in (1/2, 1)$ and let f and σ satisfy the above-presented conditions. Then, for every initial condition, the solution to stochastic differential equations converges toward a unique stationary solution in the total variation norm. The difference between stationary solution is bounded by $C_\gamma t^{-\gamma}$ for large t for every $\gamma < \frac{1}{8}$.

The uniqueness of the stationary solution pertains to the class of stationary solution adapted to the natural filtration induced by the two-sided fractional Brownian motion that drives the equation. To analyse stock market, it is going to be assumed that the movement is based only on past information, not the future ones. If the movement would be based on future solutions, there would be in theory other stationary solutions. Analysing the case of $H = \frac{1}{2}$, it is well known that the convergence to the stationary state is exponential. The total variation refers between measures on the space of paths. A stock market can be considered as a system driven by Markovian noise. Only complete, metrizable and separate space will be considered. Because processes are non-Markovian, it is necessary to know its past state, not only the present one. Markov process is the defined in the following manner.

Let $(\Omega, \mathcal{F}, \mathcal{P})$ be a probability space with a filtration $(\mathcal{F}_s, s \in I)$, for some (totally ordered) index set I; and let $(S, \mathcal{S})$ be a measure space. An S-valued stochastic process adapted to the filtration is said to possess the Markov property with respect to the $\{\mathcal{F}_t\}$ if, for each $A \in \mathcal{S}$ and each $s, t \in I$ with $s < t$,

$$\mathcal{P}(X_t \in A \mid \mathcal{F}_s) = \mathcal{P}(X_t \in A \mid X_s) \tag{7}$$

A Markov process is a stochastic process, which satisfies the Markov property with respect to its natural filtration. Non-Markovian process is defined in the same manner, but by including past state. It means that stochastic process has got memory property, because it remembers the past states.

To proceed, Hurst exponent must be defined. It is used to analyse long memory in time series. It relates to the autocorrelation time series and it is defined in terms of the asymptotic behaviour. It has the following formula:

$$\mathrm{E}\left[\frac{R(n)}{S(n)}\right] = Cn^H \text{ as } n \to \infty \tag{8}$$

where $R(n)$ is the range of the first n values, $S(n)$ is their standard deviation, $\mathrm{E}[x]$ is the expected value, n is the time span of the observation (number of data points in a time series), and C is a constant.

Hurst exponent can be estimated in various ways and it is really hard to be estimated. Some numerical methods can be used that can give approximate values to the aforementioned Hurst exponent.

To proceed with analysis, stochastic logistic differential equation with implemented fractional Brownian motion will be presented. It will be analysed and the solution will be presented.

First ordinary logistic map will be presented. It has the following form:

$$x_{n+1} = rx_n(1 - x_n) \tag{9}$$

where x_n is a number between zero and one that represents the ratio of existing population to the maximum possible population. It is obvious that for the equation where r is 1.2, x_0 has the value of 0.4. It is obvious that attractor point exists. It should also be noted that when r changes, the attractor point changes. In our case, we are speaking about dynamic r which changes continuously, so the equilibrium value changes. Delayed logistic equation looks like the following:

$$x'(t) = ax(t)(1 - x(t - \tau)) \tag{10}$$

where τ is delay parameter. For a $\tau < 1/e$, solutions are monotonic. For $1/e < at < \pi/2$, the solutions are oscillatory and asymptotically approach $x = 1$. For $a > \pi/2$, the solutions approach a limit cycle. But when considering stock market movement in China, parameter a changes constantly as well as delayed parameter. By considering the dynamic movement for values of $a = 1.74566$ and delay parameter of $\tau = 1.42783$, we get high oscillation that are continuous. If it is possible to predict the movement of a in the other words, its minimum and maximum then it is possible to predict movements, but to be more precise, we make the logistic equation stochastic one. The following representation is obtained [4]:

$$dx(t) = x(t) \times \left(1 + {}^{x(t)}\!\big/_{x_{mAX}}\right) \times \left(rdt + \sigma dw(t)\right) \tag{11}$$

where r is the growth rate dependent of t, s is the volatility (in example of the stock market) and $w(t)$ is the standard Brownian motion with Hurst exponent of 1/2. Volatility changes can be given a range, for which delayed stochastic logistic equations will be used.

It is obvious that logistic equation which was simulated achieves its equilibrium value of 4, because parameters used are $r = 1$, $x_{\max} = 4$, $x_0 = 1$, $\sigma = 1/2$. It is seen that even for oscillations of 1/2, there are many possibilities of oscillation of the value using ordinary Brownian motion.

3 Methodology and modelling

First, data for calculations were taken from the St. Louis FED and total shares prices for all shares of China were taken. By using Hurst exponent and two different packages in R, the values shown in Fig. 1 are obtained.

The following equation proves that the average Hurst 48-month value is around 0.48, so we can assume that it is close to 0.5 which mimics the ordinary Brownian motion. However, other package (FRN package R) gives the value of 0.66. Therefore, both values will be analysed to be more precise and find the equilibrium value. If Hurst exponent has the value of 0.5, then it is assumed that Chinese stock market follows ordinary Brownian motion. Taking into consideration volatility, that is taken from St. Louis FED, the following values are obtained (Fig. 2). Volatility stock price index represents the standard deviation of China stock prices in 365 days. It is obvious that stock price index is really high. Average stock price volatility is 27.5%. However, the calculated

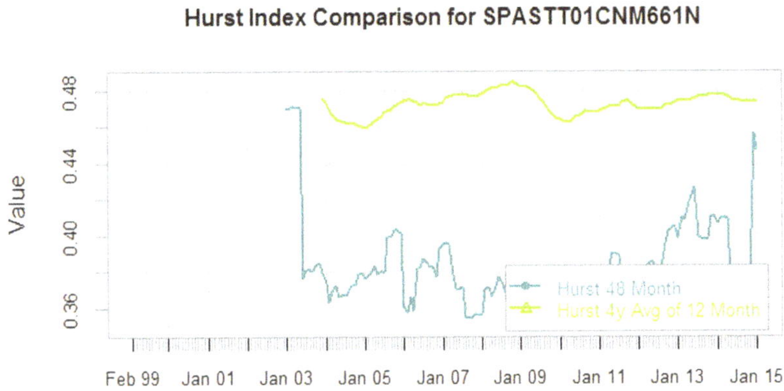

Figure 1: Hurst exponent calculated using R package.

Figure 2: World Bank, volatility of Stock Price Index for China [DDSM01CNA066NWDB], retrieved from FRED, Federal Reserve Bank of St. Louis.

value is lower than expected; therefore, we take the latest volatility level which is 30%. The abnormal standard deviations are weighted and they have been weighted to mitigate their effect. Having average stock price volatility, average growth rate should be calculated. To get the current state of index, the latest growth rate should be taken, because only the latest data should be taken into account. The growth rate taken was 29%. The stock price volatility is taken to be 30%. All the aforementioned data are taken pertaining to Shanghai stock index and its recent movements. It is higher than average and it should be taken approximately, because it is also the dynamic variable. Dynamic variables are continuously changing; therefore, only the latest values should be taken and should be changed on a constant basis to get the newest predictions.

The following values have been implemented in the stochastic logistic differential equation presented above and Fig. 3 has been obtained:

Figure 3 demonstrates the movement of Shanghai stock index for the period of April 2014 until February 2015. The average growth rate for this period was calculated. Its value is 12%. It was implemented in the above-presented stochastic differential equation and the result was obtained. Maximum value is considered to be 3500 index points, its beginning value for this period is 2100. If assumptions are correct, index will reach the aforementioned boundary and then will return to its equilibrium value which is around 3000 (according to the picture). This is expected to happen in 20 months' time and interval of equilibrium value is 3000–3100. Attention must be directed that this is a dynamic model, growth is changing constantly as well as standard deviation while beginning value and maximal value must be presumed. Although logistic growth model will always eventually attain the aforementioned value, it is important to emphasise the different paths that fluctuating stock index could follow before achieving that value. At the same time, it should be noted that this dynamical model could only work uptrending markets and it can be used to predict possible resistance levels.

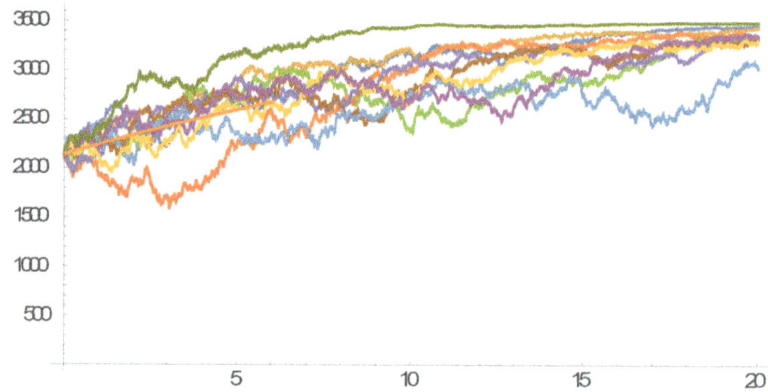

Figure 3: Stochastic logistic equation with parameters for the period April 2014 to February 2015.

Different approach when Hurst index >1/2. Now the situation will be analysed according to the Hurst index value of 0.66. The stochastic logistic delayed equation cannot be solved for the Hurst value of 0.66, because then fractional Brownian motion exists. However, according to Nguyen Tien Dung, the following theorem is proved [2] and his paper On Delayed Logistic Equation using Fractional Brownian motion.

T3. Suppose that $\mu(t)$ is an adapted stochastic process and fulfils the following conditions:

$$\sup_{0<t<T}\left(e^{\int_0^t \mu(s)ds}\right) \in L^p(\Omega) \text{ for some } p > \frac{1}{H} \tag{12}$$

$$e^{\int_0^t \mu(s)ds} \in L^q(\Omega) \text{ for some } q > 4 \tag{13}$$

$$\int_0^T \|\mu(t)\|_{1,4} \, dt < \infty \tag{14}$$

Then, the equation [15]:

$$dx_t = \mu(t) x_t dt + \sigma x_t dW_t^H \quad X_0 = x > 0 \tag{15}$$

has a unique solution in Polish space, which is given by

$$X_t = X_0 e^{\int_0^t \mu(s)ds + \sigma W_t^H} \tag{16}$$

To derive an equilibrium solution, one of the results from Ngyen Tien Dung paper will be used:

For constant $M > 0$ such that $\mu(t) \leq M$ for every $t > 0$, so we have the following result [7]:

$$X_t = X_0 e^{\int_0^t \mu(s)ds + \sigma W_t^H} \leq X_0 e^{Mt + \sigma W_t^H} \tag{17}$$

Taking volatility to be 0.1 and Hurst exponent of 0.66, W_t W_t^H is simulated and the results shown in Fig. 4 are obtained [16]:

Figure 4 demonstrates that the movement of Chinese stock market can be approximated by fractional Brownian motion [12, 14]. The next thing that must be conducted is to analyse the possible oscillations of fBM to have approximate range. After running 1000 simulations of fractional Brownian motion, the range of oscillations is from -2 to 2. Multiplying that by volatility of 0.3 and simulating exponential stochastic motion, the graph shown in Fig. 5 is obtained:

Figure 4: Comparison between total share price China and fBM.

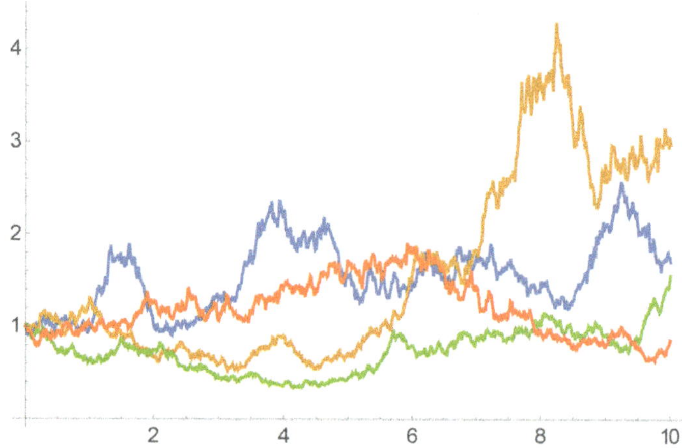

Figure 5: Exponential stochastic equation simulation.

However, the graph can be varied, but approximate solutions are there. Taking the equilibrium value and assuming that average value according to the graph is 1, we get the following solution [8, 10]:

$$X_t = X_0 e^{\int_0^t \mu(s)ds + \sigma W_t^H} \leq X_0 e \tag{18}$$

The possible solution taking into consideration the beginning value of index of 2170 and multiplying by *e*, the following result is obtained 5900 which is the highest value index obtained; therefore, it marks the index capacity. Again it is confirmed that the beginning value of index is around 2170 even for Hurst exponent of 0.66. The index defined is Shanghai stock index. Results obtained are very interesting as they mimic the movement of Shanghai stock index. The attention should be directed to dynamic hedging as X_0 is constantly changing [12]. To find and apply the following equations, first the uprising trend must be noticed. The equilibrium value will be the value that is characterised and stationary in the longer period. For example, analysing the Shanghai stock index, it is obvious that its equilibrium value was around 2170 for a longer period and after that it started increasing. Equilibrium value is dynamic and is changing continuously [12], any value that stays for a longer period can be considered equilibrium value and it is implemented in equation as X_0. This is only valid if calculated Hurst exponent is bigger than 0.5.

4 Summary

Having analysed the Shanghai stock index and Chinese stock market, it has been concluded that its value follows fractional Brownian motion. By implementing stochastic logistic differential equation and calculating Hurst exponent, it has been concluded that it is fairly difficult to predict future movement of stock market in China. To have the right prediction, growth rate that is dynamic and volatility that is also dynamical must be calculated continuously. At the same time, the beginning value as well as the ending value of an index must be assumed. Although this is really hard, if the right values are assumed and stochastic logistic equation is used, stationary state can be found. The following statement is proved by the theorem about stationary states given in the background chapter. The following strategy can be used as the trading strategy although there is still a lot of space for further research.

References

[1] Hairer, M., Ergodicity of stochastic differential equations driven by fractional Brownian motion. *Annals of Probability*, **33**, pp. 703–758, 2003.

[2] Dung, N.T., On delayed logistic equation driven by fractional Brownian motion, **7(3)**, pp. 2–7, 2011.

[3] Carmona, P., Coutin, L. & Montseny, G., Stochastic integration with respect to fractional Brownian motion. *Annales de l'Institut Henri Poincare (B) Probability and Statistics*, **39(1)**, pp. 27–68, 2003.

[4] Guillouzic, S., L'Heureux, I. & Longtin, A., Small delay approximation of stochastic delay differential equations, *Physical Review E*, **59**, pp. 3970–3982, 1999.American Physiological Society, Oxford.

[5] Dung, N.T., A class of fractional stochastic differential equations. *Vietnam Journal of Mathematics*, **36(3)**, pp. 271–279, 2009.

[6] Ferrante, M. & Rovira, C. Stochastic delay differential equations driven by fractional Brownian motion with Hurst parameter H > 1\2. *Bernoulli*, **12(1)**, pp. 85–100, 2006.

[7] Ferrante, M. & Rovira, C., Convergence of delay differential equations driven by fractional Brownian motion. *Journal of Evolution Equations*, **10(4)**, pp. 761–783, 2010.

[8] Dung, N.T., Semimartingale approximation of Fractional Brownian motion and its applications. *Computers and Mathematics with Applications*, **61**, pp. 1844–1854, 2011.

[9] Coutin, L., An introduction to stochastic calculus with respect to fractional Brownian motion, In *Seminaire de Probabilities XL*, Springer Verlag: Berlin, Heidelberg, pp. 3–65, 2007.

[10] Dung, N.T. & Thao, T.H., An approximate approach to fractional stochastic integration and its application. *Brazilian Journal of Probability and Statistics*, **24(1)**, pp. 57–67, 2010.

[11] Tiele, T.A., Multiscaling and stock market efficiency in China, *Review of Pacific Basin Financial Markets and Policies*, **7(4)**, pp. 1–22, 2014.

[12] Chiang, T.C., Lao, L. & Xue, Q., Empirical evidence of co-movements between China and global stock markets. *26th Australasian Finance and Banking Conference*, http://papers.ssrn.com/sol3/papers.cfm?abstract_id=2315040, 2012.

[13] Cont, R., Long range dependence in financial markets. Centre for financial mathematics, Ecole Polytechnique, France, http://www.proba.jussieu.fr/pageperso/ramacont/papers/FE05.pdf, 2011.
[14] Chen, L., On the chaotic dynamic analysis of Chinese stock market, Conference for Young Computer Scientists, IEEE publisher, Hunan, 2012.
[15] Nualart, D., Differential equations driven by fractional Brownian motion. *Collectanea Mathematica*, **53(1)**, pp. 55–81, 2002.
[16] Federal Reserve Bank of St. Louis, https://www.stlouisfed.org/.

Effects of mobile application to the public transportation and future editing: Istanbul case

I. Bektas
*Department of Transportation Planning,
General Directorate of Istanbul Electricity, Turkey*

Abstract

It has been possible to see the effects of vertiginous enhancements in every sector by the side of technologic developments. Especially in communication sector, these developments have been seen in the form of paradigms. Wrapping of fibre optic network cables around the world quickly, increasing the number of communication satellites, and internet becoming a normal part of life have been made easier for people's communications. However, developments in communication tools have fostered to change not only human behaviours but also organization business processes. Organizations had begun to change their organizational structure very fast synchronously with these developments from approach of total quality, strategic planning to customer centric approach. Nowadays, thanks to smart mobile phones, organizations can get in touch with their customers more easily than ever, they can understand customer's needs, even needs that have not emerged yet, and in the direction of these needs they carry out R&D processes and new product development processes. Technologic developments in communication sector had changed not only customer centric business processes but also internal transactions of organizations, and also had ensured serious increase in productivity. In this paper, development of technologies has been evaluated and has been examined in the dimension of public transportation planning, and clues have been searched for question of how will be future by giving a place to IETTs' (Istanbul Electricity, Tunnel and Tramway, Istanbul Public Transportation Authority) studies in these fields.
Keywords: analytics, IETT, public transportation planning, smart application, smart phone, web.

WIT Transactions on State of the Art in Science and Engineering, Vol 90, © 2016 WIT Press
www.witpress.com, ISSN 1755-8336 (on-line)
doi:10.2495/978-1-78466-155-7/015

1 Introduction

The rapid development and use of communication technologies and, at the same rate, involvement between people, easy accessibility to these instruments, especially vertiginous enhancements in web application, have fostered big change in the dimension of organization business processes in the past three decades. Not only integrated innovative leaps in web and communication technologies have generated new products such as smart mobile phones, social media platforms that people get connected easily, but also these improvements have generated big data that bring complexity for analysing these huge amount of data and converting them into information.

The evolution of production-oriented organizations is reserved briefly in three periods. First one was the production-oriented period before 1970s. Demand was high but due to lack of producers supply was low. Customers had to buy what producers offer them. Second one was quality product-oriented period between 1970 and 1990. In that period, the number of producers has been increased and the effect of customers has also increased. Organizations had to produce more quality products to ensure loyalty of customers. Third one was customer-oriented period after 1990s. It was not enough to produce quality product for customers. Customers have the power of choosing any product in a large scale of product supply. Even power of making organization to produce customize product as customer wish. Last period has forced organizations not only to understand customer needs but also seek the way to reach customers. Transparent approach for customization (Yüksel, B. 1990) is an approach which envisages understanding the client without asking again and again, but still examining the needs of the client to interact and produce appropriate solutions to customers. However, for this approach to determine the client's needs in the 1990s was very difficult because of more limited data generated than the present. With the rapid development in Information and Communication Technology (ICT), many interactions, especially people's shopping habits, methods of communication began to change. Public and private organizations are increasing their investment in ICT. ICT has become the main element of public or private status of all business activities [1].

In this paper, information and communication developments in the field of web and smart phone technology and how to use it and work with these developments for public transportation planning and future vision for the IETT studies will be explained.

2 Customer behaviour

The theories, both Justified Action Theory and the Theory of Planned Behaviour [2–4], have been developed and widely accepted among researchers to explain the behaviour in perspective of social psychology. In psychology, the theory of planned behaviour (abbreviated as TPB) is a theory, which links beliefs and behaviour. It has been applied to studies of the relations among beliefs, attitudes, behavioural intentions, and behaviours in various fields such as advertising, public relations, advertising campaigns, and health care. With this perspective, organizations realized marketing activities switching to customers centralized approach, understanding the customer, listening to even planning production processes with customers to produce and working on customer satisfaction for performing the highest level with the help of ICT.

Development of ICTs has facilitated the ease of gathering customer data. With the help of such systems on the internet as online customer tracking (OCT), organizations are able to make sense of these client data on an individual basis by following the customer communication with any online device and present suitable offers for customers with this evaluated information (Fig. 1).

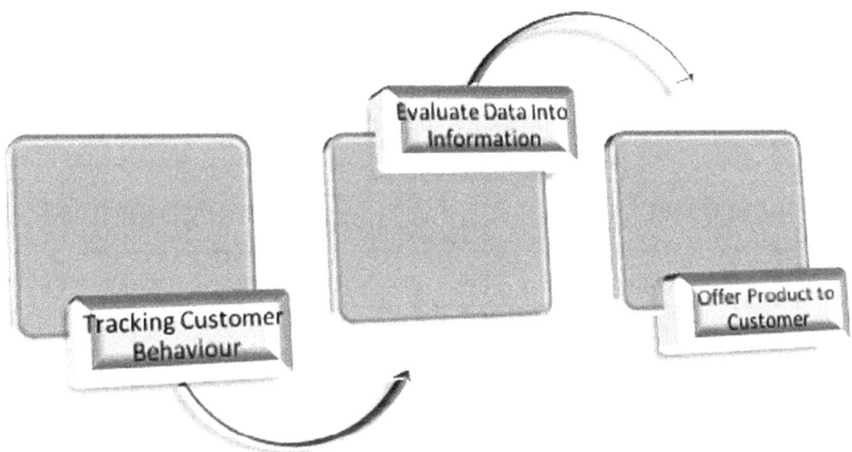

Figure 1: Marketing strategy based on customer tracking.

3 Development of web version

Between 1989 and 1999 years, with Web 1.0 people cannot just unilaterally make no interaction. It was an environment where the user is in a passive state by searching on the internet. During the years 1999–2009, people started interacting. People can share their thoughts on the site and have developed their own sites and blogs with Web 2.0. These activities have led to the increasing emergence of social media, which is quite common nowadays.

Social online platforms, in any focused topic, have endorsed saying 'All the people, when they combined, they know more than a few people who specialize in that area regarding any issues' [5].

When surfing on the internet with this interaction, people have to leave their own distinctive digital footprints. This digital track, relevant with the people themselves, almost include some of the details such as time and logical attitude to their emotional state, including venue information, entertainment understanding to the political approach and a lot more constructs as 'Individuals Digital Information Map'.

4 Development of smart phone technology

Smart phone, a product of the world's PDA (personal digital assistant), is an advanced mobile communication, the features of the device being designed as a pocket computer with the addition of classic features provided by the mobile phone. Smart phones are equipped with mobile operating systems. In this manner, they can be used in a more active way for many different purposes and applications can be found for almost every need.
This situation has resulted in the formation of an information and communication platform that is more sophisticated and generate more data and individuals with the rapid increase in mobile phones development and the dizzying advances in smartphone penetration and usage. Smart phone usage rates in the world are shown in Table 1 [6].

Table 1: World smart phone number of connections per 100 citizens.

Country rank	Connections/100 citizens	Population
1. Hong Kong	236.8	7,219,700
2. Panama	202.5	3,405,813
3. Saudi Arabia	169.5	27,137,000
4. Lithuanian	167.1	2,955,986
5. Estonia	159.9	1,294,486
6. Russia	155.5	142,905,200
7. Singapore	151.8	5,312,400
8. Italy	147.4	60,090,400
9. Malaysia	143.8	28,250,000
10. Argentina	141.34	40,134,425
11. Bulgaria	140.2	7,600,000
12. United Arab Emirates	139.6	8,264,070
13. Brazil	136.45	201,032,714
14. Australia	133	22,700,000
15. Germany	130.1	81,882,342
50. Turkey	89.9	75,627,384

Our Mobile Planet Turkey has conducted a survey with 1,000 private smartphone users, who use the internet in general, participated in the study. The survey indicates the context of the place of smartphones use for entire 2013 [7]. As can be seen in the graph, the user uses the smart phone in any environment. There is an increased smartphone usage rates in locations where user's mobility duration is stable. In Fig. 2, the context of the place of smartphones use is summarized.

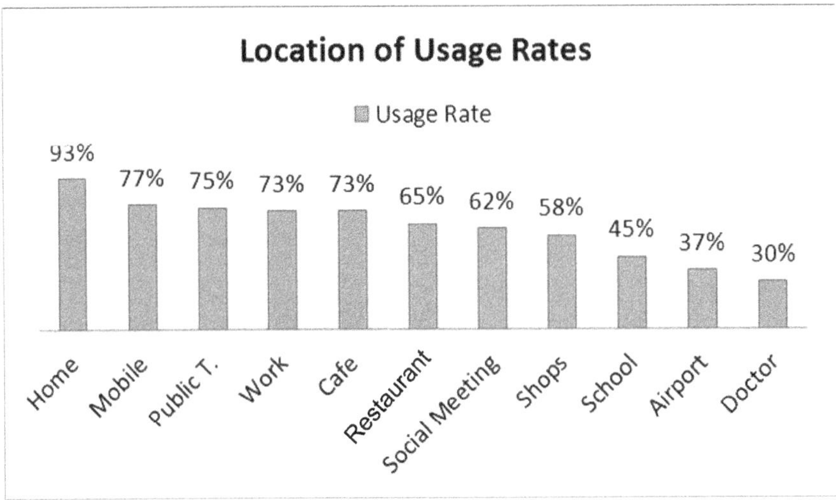

Figure 2: Location of usage rates.

5 Public transportation perspective

Every year seven cities are added, equivalent to New York, in the world. Becoming the cities as the centre of attraction has increased the gravity force from country to cities and has led to overcrowding of the urban population. Today, serious transportation problems have manifested in the big cities because of increasing urban population. The transportation problems delay or make an unattainable goal of the cities and the countries sustainable economic development.

These dizzying advances in Information and Communication Technologies have contributed to public transportation, which is a chronic problem with increasing urban population and private car ownership. The most innovative example is the electronic cards that are used for public transportation payment systems by many cities and countries in the world. Electronic cards have reduced the loss of revenue as well as passenger waiting time. Also, it provided making the public transport operators planning models and analysis more quickly with the passenger boarding data.

In Fig. 2, the statistical research that indicates that 75% of the users use smart phones during public transport should be underlined. Ensuring users to conduct the relocation process except their own mechanical activity reveals an interesting situation. Public transport users use smart phones for a variety of social media, game or entertainment applications when they are on the move in public transportation vehicles. This situation has led to the orientation of the public transport authorities to mobile applications. Mobiett is one of the most potent examples of these applications [8].

All public transport authorities use data that obtained from past electronic cards data for the assessment of basic planning principles, to make operational tactical and strategic plans in the world.

With rapidly increasing use of smart phones, internet users have left digital trail in the virtual environments that are began to be matched with real spatial points. Mobility of people has become well monitored together with the use of social media and Geographic Information System (Foursquare, etc.). Mobility in these researches is emerging to be specific to each person. This technological integration with public transport is seen as closely related with the following three perspectives [9]:

1. Map, place and route location.
2. Navigation, mobility and relocation.
3. Behaviour, reason of mobility and mode of mobility.

In public transportation planning, the first two perspectives are the often used perspectives of planning. But the third perspective is not used. Instead of that, past trends are used for public transportation planning (Fig. 3).

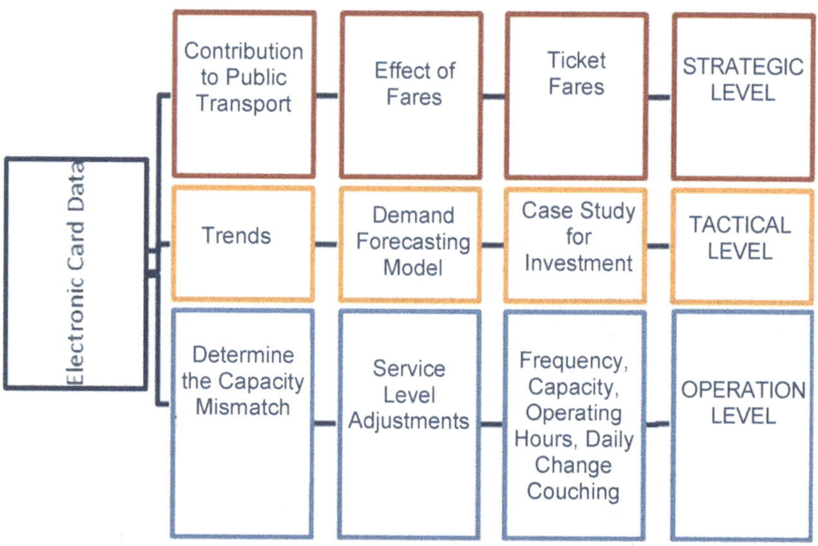

Figure 3: Three-level planning made with electronic card trip data.

IETT has started 'Public Transportation Navigation Project'. It is based on a system that integrates smartcard data and mobile application on a website–based platform. In Istanbul, population is over 14 million and the numbers of smartcards in use are over 16 million, and this number is increasing very fast as shown in Fig. 4. The number of usage is over 6 million on daily basis.

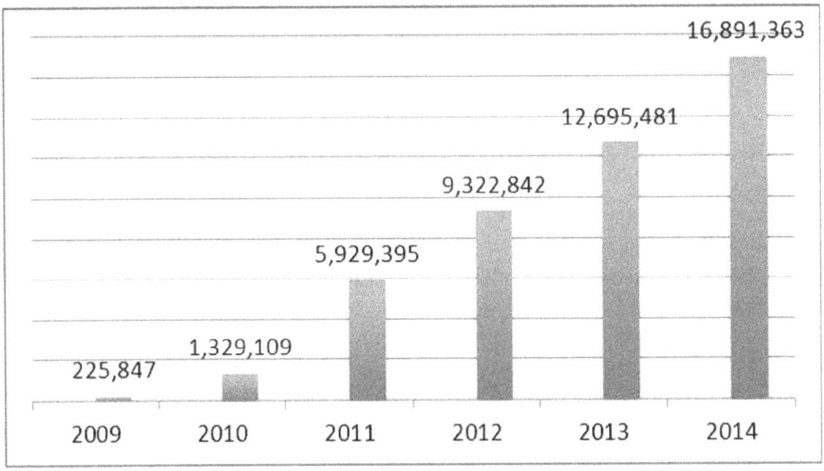

Figure 4: Numbers of Istanbulcards (Smardcard) in use by years.

Istanbulcard data are used for ridership analyses on public transportation infrastructure by IETT. It is possible to track every Istanbul card ID on the route. Tap-on is easy to track, for tap-off actions some analyses have to be made, but not so that difficult, and a sample calculated boarding and alighting is shown in Table 2.

Table 2: Sample for number of passenger from origin to destination on the basis of bus stops.

Alighting					
	Code of bus stops	A0284	A1158	A0285	A0286
Boarding	A0340	0	2	0	0
	A0339	1	1	1	0
	A0338	2	0	0	0
	A0337	1	0	0	0
	A0336	0	0	0	1
	A0334	0	0	0	0
	A0333	0	0	0	0
	A0332	0	1	1	0
	A1911	1	2	4	1
	A0754	0	0	0	0
	A3301	1	0	0	0
	A3302	0	0	1	0
	A3303	0	0	0	0

IETT measures reliability punctuality performance with AVL (automatic vehicle location) system that enables to track vehicle in real time.

IETT aims to construct an online platform, which will gather and merge all types of data that it have. All Istanbulcard holders will be able to register this platform by smartcard ID. They will able to form their profile and will be provided all kinds of Istanbulcard transactions such as online payment and ordering new personalized Isatnbulcard, which are now available by Istanbulcard centres and prepayment machines. Platform is not limited for these kinds of functions. Platform will be supported by mobile application and Istanbulcard data. With data fusion of two big data, Istanbulcard data and AVL data, users will be able to see not only their journeys history on a simulation which will be formed with real personalized mobility data, but also real-time vehicle locations on the route, and they will be able to share that simulation on social media platforms. Their journey history will be recorded for scoring trips activity like scores according to transportation mode and contribution to sustainable transportation improvements.

Interaction with other registered users on platform will also take part in this project. With that users will be able to share public transportation information on communities that will be open for all users according to specifications such as users on the same route or mode, very fast.

There will be three modules apart from these common functions:

- Public transportation coaching (PTC) [10].
- Gamification.

PTC module will be activated as per demand of users. If a user want to join this module, the user will give permission to IETT track him/her by mobile application and personalized Istanbulcard data on public transportation infrastructures. After analysing user's journey experience, the module will offer to user an optimum new journey plan according to constrains (comfort, travel time, cost, etc.) that are determined by the user (Fig. 5).

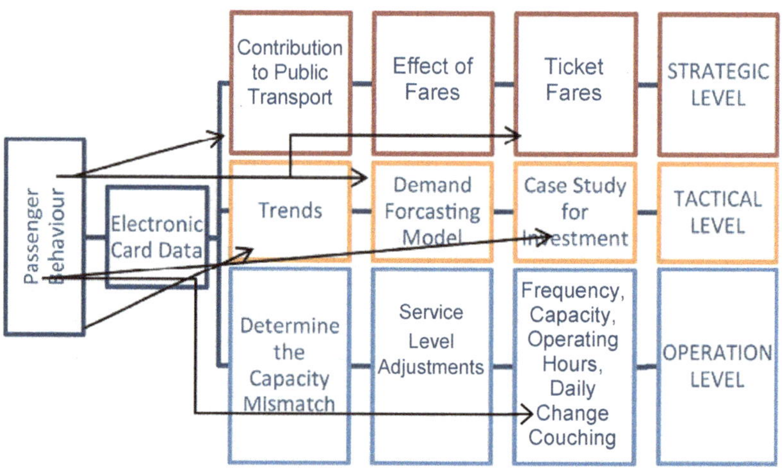

Figure 5: IETT three-level planning made with electronic card trip data and passenger behaviours.

Gamification module will be activated as per demand of users. Users will be scored on specific success missions. According to scores, users will gain ranks.

These modules will provide a unique opportunity for IETT to get in touch with its customer and understand their transportation needs, behaviours and take their contribution for planning. For example, if IETT prepare to invest on a new route one point to another, IETT will easily take passengers who use public transportation between this points, thoughts and collect demand to decide if the investment will be feasible or not in the terms of passenger's perception. That will prevent false investments or will improve sharpness of forecasts and increase customer satisfaction.

6 Conclusion and recommendation

Operational sense, public transportation couching service will direct passengers to compatible modes of public transportation.

Tactical sense: attraction centres will emerge with passenger data to be obtained easily.

Strategic sense: ticket prices can be removed from the fixed price to flexible pricing system with dynamic pricing for each individual passenger needs.

Results of 'Public Transportation Navigation Project' will be shared in the future paper.

References

[1] Hu, P.J., Chau, P.Y., Hu Sheng, O.R. & Tam, K.Y., Examining the technology acceptance model using physician acceptance of telemedicine technology. *Journal of Management Information Systems*, **16(2)**, pp. 91–112, 1999.

[2] Ajzen, I., From intentions to actions: a theory of planned behavior. *Action Control, From Cognition to Behaviour*, eds. J. Kuhl & J. Beckman, Springer-Verlag: Berlin, Heidelberg, pp. 11–39. 1985.

[3] Ajzen, I., The theory of planned behaviour. *Organization Behaviour and Human Decision Process*, **50**, pp. 179–211, 1991.

[4] Fishbein, M. & Ajzen, I., *Belief, Attitude, Intention and Behavior: An Introduction to Theory and Research*, Addison-Wesley: Reading, MA, 1975.

[5] Nash, A., Web 2.0 applications for improving public participation in transport planning, *Transportation Research Board 89th Annual Meeting Vienna Transport Strategies*, Austria, 2009.

[6] List of countries by number of mobile phones in use. http://en.wikipedia.org/wiki/List_of_countries_by_number_of_mobile_phones_in_use (accessed 28 April 2015).

[7] Our Mobile Planet, Understanding Mobile Consumer Turkey Report. http://services.google.com/fh/files/misc/omp-2013-tr-local.pdf (accessed 28 April 2015).

[8] http://www.iett.gov.tr/tr/main/news/mobiett-ile-artik-her-durak-akilli/1397

[9] Goodchild., M.F., GIS and transportation: status and challenges. *GeoInformatica*, **4(2)**, pp. 127–139, 2000.
[10] Bektas,, I˙, *TRANSIST 2014 Papers*, pp. 252–258, 2014.

Section 3
Applications in various fields

Supporting sensemaking to deal with organizational complexity

S. Jantunen
Lappeenranta University of Technology, Finland

Abstract

Many organizations have in recent years experienced increased inadequacy of their management practices to deal with their complex and turbulent work-related challenges. Failing to recognize the complexity in their problems, organizations have persisted in applying inappropriate methods and tools to them. When the work-related problems are ambiguous, complex, and constrained, business processes tuned for decision-making have begun to degrade providing a false sense of security based on unjustified numerical certainty. Examining these challenges in the light of complexity theory could potentially help organizations to deal with their challenges in a more effective way. For understanding complex and ambiguous issues, people must interact, discuss, and share their different perspectives and interpretations. Existing literature recognizes sensemaking as one way for group members to understand and talk about complexity. Unfortunately, little has yet been reported on exactly how the properties of sensemaking can be materialized in practice. This paper discusses sensemaking as a potential framework to deal with complexity. The paper (1) describes on a practical level interventions taken in one case company to support sensemaking, (2) illustrates the creation of different plausible interpretations of what essentially describes the case company's current environment, (3) describes how the creation of shared understanding is currently being shaped into a commitment towards action, and (4) reflects the findings and experiences with the existing knowledge on complexity and sensemaking.
Keywords: business processes, complexity, sensemaking.

WIT Transactions on State of the Art in Science and Engineering, Vol 90, © 2016 WIT Press
www.witpress.com, ISSN 1755-8336 (on-line)
doi:10.2495/978-1-78466-155-7/016

1 Introduction

During the recent years, we have witnessed rapid development of information technology resulting with increased interconnectedness, transparency, empowerment of individuals, speed of transactions, and decreased cost of information [1]. These are all trends that are likely to introduce complexity to the way organizations manage their businesses. Advances in technology have led many firms to the point that innovations cannot be accomplished any more by individuals acting independently [2, 3]. Furthermore, continuously deepening globalization has forced firms to find their competitive advantage from their flexibility, agility, capability to innovate, and ability to understand the customer needs [4]. Consequently, firms have been driven to a situation in which they need to be able to collaborate effectively with other firms, whether suppliers, customers, or competitors, and with other non-commercial knowledge-generating organizations.

When the business contexts have turned out to be ambiguous, complex, and constrained, processes tuned for decision-making have begun to degrade providing a false sense of security based on unjustified numerical certainty. Failing to recognize the complex and turbulent dynamics in problems at hand, firms persist in applying inappropriate methods and tools to them [5]. Examining the work-related challenges in the light of complexity theory could potentially help organizations to deal with their challenges in a more effective way [6].

For understanding complex and ambiguous issues, people must interact, discuss, and share their different perspectives and interpretations. Existing literature recognizes sensemaking [7] as one way for group members to understand and talk about complexity [8]. Unfortunately, little has yet been reported on exactly how the properties of sensemaking can be materialized in practice.

This paper discusses sensemaking as a potential framework to deal with complexity. The paper (1) describes on a practical-level interventions taken in one case company to support sensemaking, (2) illustrates the creation of different plausible interpretations of what essentially describes the case company's current environment, (3) describes how the creation of shared understanding is currently being shaped into a commitment towards action, and (4) reflects the findings and experiences with the existing knowledge of complexity and sensemaking.

The remainder of this paper is organized as follows. Section 2 introduces the framework of sensemaking in greater detail, Section 3 describes the empirical context including the case company and the research project. Section 4 describes the research methodology, data gathering, and empirical results. Section 5 discusses the findings and reflects the experiences to the literature of complexity and sensemaking. Finally, Section 6 concludes the paper.

2 Sensemaking

Sensemaking, a framework proposed by Weick [7], seeks to explain how people structure unknown so as to be able to act in it. Sensemaking enables us to turn the ongoing complexity of the world into 'a situation that is comprehended explicitly in words and that serves as a springboard into action' [9]. In more concrete terms, sensemaking involves (1) coming up with a plausible understanding – a map of a shifting world; (2) testing this map with others through data collection, action, and conversation; and then (3) refining, or abandoning, the map depending on how credible it is [10].

Sensemaking is an ongoing process that is triggered when the current state of the world is perceived to be different from the expected state of the world, or when there is no obvious way to engage the world [9]. At its most basic, sensemaking is about understanding how different meanings are assigned to the same event [11]. According to Weick [7], at least seven distinguishing characteristics set sensemaking apart from other explanatory processes such as understanding, interpretation, and attribution. Sensemaking can be understood as a process that is [7]:

1. *Grounded in identity construction.* The sensemaker is 'an ongoing puzzle undergoing continual redefinition, coincident with presenting some self to others and trying to decide which self is appropriate' [7, p. 20]. The definition of what is 'out there' depends on individual's perception of his or her identity. One's perception of identity also affects the actions one makes. Weick [7, p. 23] describes several points worth noting about identity. First, controlled, intentional sensemaking is triggered by a failure to confirm one's self. Second, sensemaking occurs in the service of maintaining a consistent, positive self-conception. Third, people learn about their identities by projecting them into an environment and observing the consequences. Fourth, people simultaneously try to shape and react to the environments they face. There is a complex mixture of proaction and reaction, and this complexity is commonplace in sensemaking. Fifth, the idea that sensemaking is self-referential suggests that self, rather than the environment, may be the text in need of interpretation.
2. *Retrospective.* The world people perceive is in reality a past world. However, whatever is occurring at the moment will influence what is discovered when people glance backward [7, p. 26]. Because events to be interpreted have already elapsed, anything that affects remembering will also affect the sense that is made of those memories. Elapsed experience appears to be equivocal, not because it makes no sense at all, but because it makes many different kinds of sense [7, p. 27]. The important point is that retrospective sensemaking is an activity in which many possible meanings may need to be synthesized. In such situation, what is important is clarity on values that indicate what is important in elapsed experience, which finally gives some sense of what that elapsed experience means [7, p. 28].
3. *Enactive of sensible environments.* People often produce part of the environment they face, while the environment partly defines who the people are. There is *not* some kind of monolithic, singular, fixed environment that exists detached from and external to people. Instead,

the people are very much a part of their own environments. They act, and in doing so create the materials that become the constraints and opportunities they face [7, p. 31].

4. *Social.* Individual's thinking and social activities interact closely with each other. Sensemaking is never solitary because what a person does internally is contingent on others [7, p. 40]. One has to fit one's own line of activity in some manner to the action of others.
5. *Ongoing.* Sensemaking never starts or stops. People find themselves thrown into ongoing situations. To understand sensemaking is to be sensitive to the ways in which people chop moments out of continuous flows and extract cues from those moments [7, p. 43].
6. *Focused on and by extracted cues.* Extracted cues are simple, familiar structures that are seeds from which people develop a larger sense of what may be occurring [7, p. 50]. People in organizations are in different locations and are familiar with different domains, which mean that they have different interpretations of common events [7, p. 53].
7. *Driven by plausibility rather than accuracy.* Accuracy is meaningless when used to describe a filtered sense of the present, linked with a reconstruction of the past that has been edited in hindsight [7, p. 57]. Instead of accuracy, we need something that: (1) preserves plausibility and coherence, (2) is reasonable and memorable, (3) embodies past experience and expectations, (4) resonates with others, (5) can be constructed retrospectively but also can be used prospectively, (6) captures both feeling and though, (7) allows for embellishment to fit current oddities, and (8) is fun to construct. In short, what is necessary in sensemaking is a good story [7, p. 60].

3 The case company and the research project

Although sensemaking has been found to be a potential way to deal with complexity, little has yet been written how the seven properties of sensemaking [7] (Section 2) can be materialized in practice. Furthermore, little empirical reports exist describing experiences of conducting sensemaking activities within organizations. This paper seeks to contribute to these topics by describing empirical work within one case company. This section describes the case company in greater detail, along with the research project within which the empirical activities have been conducted.

Suomen Pienyrittäjäin Mainostoimisto (SPYM) is an advertisement company that offers services for business development, website design/development, graphical design, media advertising, digital marketing, communication, and content creation. SPYM was founded in 2008 in Jyväskylä, Finland. The company started their business by offering advertisement services to small businesses. Since its establishment, SPYM has systematically expanded its business. Today, SPYM employs 14 persons in the cities of Jyväskylä and Helsinki. In addition, SPYM utilizes several sub-contractors for a wide range of activities. In recent years, SPYM has started to serve increasingly larger customers while still serving the small businesses. This seems to have started to introduce new sources of complexity to the way SPYM operates. Larger customers seem to expect more disciplined collaboration compared to the small businesses who prefer hands on attitude and quick responses. Balancing between planned and *ad hoc* modes of working appears to have led to confusion, inefficiency, and eventually exhaustion of SPYM's employees. It has become evident that SPYM needs to find new ways of working in order to be able to cope with new sources of complexity in their business environment.

The necessity to find new ways of working has been the primary motivation for SPYM to join a recently established Organization 2.0 research project. The project, funded by the Finnish Funding Agency for Innovation (TEKES), is planned to be conducted from August 1, 2013 to December 31, 2015 in close collaboration with industrial partners by researchers originating from three Finnish research institutions: JAMK University of Applied Sciences, Lappeenranta University of Technology, and VTT Technical Research Centre of Finland. The Organization 2.0 project seeks to understand better *what kinds of management practices would enhance firm's capabilities to cope with constant complexity and change* (Research question 1) and *how the novel management practices affect employees' job satisfaction and meaningfulness of work* (Research question 2). To this end, research is conducted in two distinct work packages. The first work package identifies and investigates organizations that have developed novel and alternative ways to deal with their work related challenges. This work package seeks to (1) identify management practices that have proved to be effective and (2) explain the success of these practices by reflecting the findings with scientific knowledge. The research collaboration with SPYM is part of the second work package, which seeks to (1) develop novel management practices on the basis of scientific knowledge and (2) experiment and evaluate their effectiveness in real-life contexts.

4 Research method and results

The study in SPYM has followed action research methodology, which has been found to provide a valid methodological approach to the study of complexity [12]. Although action research is commonly credited to be originated by the work of social psychologist Kurt Lewin in the first half of the twentieth century, it is important to be aware that the methodology has since then evolved rather independently in different disciplines. Considering the history of action research, it is not surprising that there are now many different definitions and approaches for it [13]. Being aware of these differences is important, because different action research approaches have differing philosophical underpinnings and should hence be conducted and evaluated according with different criteria [14, 15].

Our action research approach at SPYM resembles *emancipatory action research* that has its philosophical roots in critical theory. This form of action research seeks to 'assist practitioners in identifying and making explicit fundamental problems by first raising their collective consciousness' [16]. Newton and Burgess [15] have suggested that emancipatory action research should primarily be evaluated in terms of what Anderson and Herr [14] refer as the *democratic* and *catalytic* validities. Democratic validity is concerned with 'the extent to which research is done in collaboration with all parties who have a stake in the problem under investigation' [14]. Catalytic validity refers to the degree to which the research process have energized participants toward knowing reality in order to transform it [14].

Our attempt to conduct emancipatory action research has resulted with interventions, which have been designed to encourage the relevant stakeholders to understand their problematic situation themselves. In reality, these interventions have taken the form of workshops that have been organized at the premises of the case company. An overview of the workshops that have been organized to this point is presented in Table 1.

Table 1: Organized workshops, participants, and gathered data.

Workshop 1: Exploration of current situation (May 15, 2014) Participants: (1) Managing Director, (2) Production Manager, (3) Graphical Designer
• 75 identified things that are considered good in the current work environment • 60 identified things that a participant would wish to see improved in their work environment • Participant-specific views of ten most essential good things (in the order of significance with written justification for each selection) • Participant-specific views of 10 most significant things that need improvement (in the order of significance with written justification for each selection) • Recorded discussions (54 min)
Workshop 2: Understanding how salient issues are related to each other (October 7, 2014) Participants: (1) Managing Director, (2) Production Manager
• Individually identified and documented relationships between emerged salient themes • Recorded discussions (60 min)
Workshop 3: Towards shared understanding and common goals (December 3, 2014) Participants: (1) Managing Director, (2) Production Manager
• Collectively developed understanding of current situation • Collectively developed agreement of common goals • Recorded discussions (45 min)

The first workshop was organized in May 15, 2014, with the objective to explore the context that SPYM is currently operating. To this end, three of SPYM's key persons including Managing Director, Production Manager, and Graphical Designer were first asked to brainstorm (1) things that they considered good in their current operating environment and then (2) things that they would wish to see improved. The brainstorming was conducted electronically by using a freely available software service, padlet.com. In practice, the participants used their personal laptops to input their ideas into a shared worksheet that displayed all ideas real time. These activities resulted with two separate worksheets with 75 things that were considered good in SPYM's current work environment, and 60 things that were wished to be improved in the current work environment. After the brainstorming session, the participants were asked to select 10 most significant things that they consider good in their work environment and 10 most significant things that they would wish to see improved. The participants were also asked to write down justifications for their selections. The participants were then asked to present their top 10 lists. These presentations were recorded.

After the workshop, the researcher exported all data from padlet.com to a spreadsheet application in such way that each of the participants' responses was represented as an individual text box. Each text box was also color-coded in order to illustrate which of the questions the response was answering to. Green text box represented things that were considered good. Red text box represented things that were wished to be improved. The analysis of gathered responses proceeded from here visually by dragging the textboxes so that similar responses were moved close to each other. This step of analysis resulted with an interpretation of gathered data, described as a map of textbox clusters with similar statements next to each other. Each of the clusters was then labeled with a name that resembled a theme that emerged from the gathered data. The emerged themes, translated from Finnish to English, are presented in Fig. 1.

The second workshop was organized in October 7, 2014. In this workshop, the emerged themes were presented and discussed. Both Managing Director and Production Manager were then given an A3 paper, similar with Fig. 1. The participants were asked to individually draw connections between the themes and describe how the themes are related with each other. The participants were then asked to present their work. These presentations were recorded.

Based on the gathered data from the second workshop, the researcher created two diagrams, each of which described individual views of how essential themes were related with each other. These diagrams were then further interpreted in order to create a simplified understanding of what could be the most salient themes to be improved and what kinds of actions could likely to facilitate such improvement. An analysis of the first individual view resulted with two different simplified interpretations (Fig. 2). First of the interpretations argued that *exhaustion* of personnel can be managed: (1) by improving *communications*, which can help to minimize *haste and ad hoc assignments* and (2) through *management and leadership practices* by improving *feedback* mechanisms,

Figure 1: Essential themes that emerged from the first workshop. In themes displayed with green background, a majority of the brainstormed items contained things that were considered good. Themes displayed with yellow background contained equal amounts of items that were good and items that needed improvement. Themes displayed with red background contained largely things that needed improvement. The layout of the figure has been modified for the purposes of this paper.

Figure 2: Two simplified interpretations derived from one individual view describing how emerged themes are related with each other.

clarification of *roles*, and *development of skills*. The second interpretation argued that the improvement in *communications* practices could enable improvements in *scheduling* and mitigate the challenges with *haste and ad hoc assignments*.

The analysis of the other individual view resulted with interpretation arguing that that the ultimate goal of the firm is profitability that can be defined in terms of quality and efficiency (Fig. 3). According to this interpretation, profitability is more likely to be reached if there are right kinds of conditions in terms of facilities, personnel, and work practices. In this interpretation, *communication* was seen as a key enabler for personnel to develop effective work practices to pursue for profitability.

At this point, the researcher also analyzed top 10 lists that were individually created in the first workshop. The results of this analysis highlighted the clear differences in individual thinking regarding what was considered good in SPYM's working environment and what was wished to be improved (Fig. 4).

All analysis results were presented and discussed in the third workshop on December 3, 2014. The purpose of this workshop was to come up with a shared view of SPYM's current working environment and to decide what will be the focus of forthcoming development activities. The presented research results (including Figs 1–4) met acceptance among the participants. These findings turned out to be well in line with the recent discussions of SPYM's management. Consequently, SPYM's management outlined in the workshop that, although profitability is the ultimate goal of the company, improvements in profitability are likely to be obtained by improving: (1) *communications* practices and (2) ways to clarify the *roles* of SPYM's personnel.

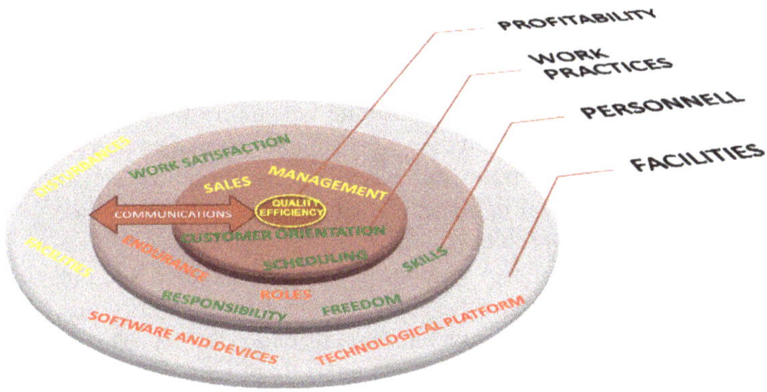

Figure 3: A simplified interpretation derived from the other individual view describing how emerged themes are related to each other.

Figure 4: Analysis of personal views of what is good in the case company's working environment and what was wished to be improved.

Now that communications and roles were selected as two primary themes to concentrate on, the focus of research collaboration started to shift from the development of shared understanding towards action. The management of SPYM agreed that a new workshop shall be organized in which all employees of SPYM shall participate. This workshop seeks to (1) further discuss and adjust the interpretation of SPYM's environment, (2) gain the acceptance and the involvement of SPYM's employees and (3) initiate the collective dialog for developing ways to improve communications and to clarify roles within SPYM.

5 Discussion

A brief literature reveals that complexity thinking, sensemaking, and emancipatory action research all seek to explain and help us to understand the nature of reality so that we can deal with it in a more effective way. While complexity thinking argues that 'examination is merely a starting point and not a mapping, and that social systems need to be studied in their own right as complex social systems' [6], sensemaking viewpoint believes that 'sensemaking is not about truth and getting it right. Instead, it is about continued redrafting of an emerging story so that it becomes more comprehensive, incorporates more of the observed data, and is more resilient in the face of criticism' [9]. These kinds of arguments appear to be well aligned with the objective of emancipatory action research that seeks to 'increase the closeness between the actual problems encountered by practitioners in a specific setting and the kind of theory used to explain and resolve those problems' [16].

When we started to collaborate with SPYM, we adopted sensemaking as a theoretical framework to guide the design of the conducted workshops. Hence, in this section we reflect our interventions through the seven properties of sensemaking (Section 2). In our view, the analysis of participant-specific top 10 lists has illustrated how perceptions of stakeholders are clearly related to their professional identities (*1. grounded in identity construction*). Each participant had a distinct view of what was good in SPYM's work environment and what need to be improved. We have made an effort to make these differences in viewpoints explicit, so that the participants can reflect these findings against their past experience (*2. retrospective*) and investigate differing viewpoints in their own work environments (*3. enactive of sensible environments*). Throughout our collaboration with SPYM, we have encouraged for participatory approaches (*4. social*). We have also highlighted that our results are merely interpretations of current situation and that what is important is a change in attitude towards continuous adaptation (*5. ongoing*). Our interventions have been design to identify real issues in their work environment (*6. focused on extracted cues*) in order to build a more holistic explanation of what may be the story explaining these findings (*7. driven by plausibility rather than accuracy*).

6 Conclusions

This paper illustrated one way of supporting sensemaking in practice. Our reflection of the conducted activities within the case company suggests that sensemaking can be seen as one potential way of thinking to deal with complexity. The gathered feedback from the workshop participants also supports this argument. At the end of each workshop, we have asked the participants to share their thoughts of the conducted activities. These responses indicate that the participants have seen value in the process and potential in the forthcoming workshops.

> What we did in the workshop was something that I could have personally addressed even for the whole day. That is, to think deeper about what is good in our working environments and how we could improve. I feel that what we achieved in this workshop was very a quickly accomplished picture about the essential themes of our working environment. That is, what is good about it and what needs to be improved. We also had a good discussion about these topics and learned from different perspectives. It is good to continue from here and think how we can develop our practices. This was interesting. *Managing Director after the first workshop*

> I have recently spent quite a lot time to think about how to improve our efficiency and how to manage our sub-contractor network. Today, I have tried to take a different angle and think in terms of our daily routines. I was able to see quite quickly that communications is one of the essential issues to deal with in future. *Production Manager after the first workshop*

> This was very interesting. I think we took just the right kind of step to the next level. *Production Manager after the second workshop*

> It was nice to look at the documentation of the previous workshops and see how things have already changed since then. What I would wish from the future is an increased understanding of how to utilize our increased understanding when developing our working environment. It would be nice to know in greater detail what precisely needs to be improved and what we want to preserve. *Managing Director after the third workshop*

Now that the case company has created a shared understanding of what aspects of work shall be improved, our focus begin to shift towards supporting the development of practical ideas that implement the commonly shared goals.

References

[1] Palmberg, K., Complex adaptive systems as metaphors for organizational management, *Learning Organization*, **16(6)**, pp. 483–490, 2009.

[2] Coombs, R. & Tetcalfe, J.S., Organizing for innovation: co-ordinating distributed innovation capabilities. *Competence, Governance, and Entrepreneurship. Advances in Economic Strategy Research*, eds. N. Foss, V. Mahnke, Oxford University Press: Oxford, pp. 209–231, 2000.

[3] Chesbrough, H., *Open Innovation. The New Imperative for Creating and Profiting From Technology*, Harvard Business School Press: Boston, MA, 2003.

[4] Alasoini, T., *Mainettaan Parempi Työ*, Taloustieto Oy: Helsinki, 2010.

[5] Conklin, J., Chapter 1: wicked problems & social complexity. *Dialogue Mapping: Building Shared Understanding of Wicked Problems*, Chichester: West Sussex, England, 2005.

[6] Mitleton-Kelly, E., The principles of complexity and enabling infrastructures, *Complex Systems and Evolutionary Perspectives on Organisations: The Application of Complexity Theory to Organisations*, ed. E. Mitleton-Kelly, Elsevier Science Ltd: Oxford, UK, pp. 23–50, 2003.

[7] Weick, K.E., *Sensemaking in Organizations*, Sage Publications: Thousand Oaks, CA, pp. 23–50, 1995.

[8] Rutledge, M., Sensemaking as a tool in working with complexity, *Organization Development Practitioner*, **41(2)**, pp. 19–24, 2009.

[9] Weick, K.E., Sutcliffe, K.M. & Obstfeld, D., Organizing and the process of sensemaking, *Organization Science*, **16(4)**, pp. 409–421, 2005.

[10] Ancona, D., Sensemaking: framing and acting in the unknown. *The Handbook for Teaching Leadership: Knowing, Doing, and Being*, eds. S. Snook, N. Nohria, R. Khurana, Sage Publications Ltd: Thousand Oaks, CA, pp. 3–21, 2012.

[11] Mills, J.H., Thurlow, A. & Mills, A.J., Making sense of sensemaking: the critical sensemaking approach. *Qualitative Research in Organizations and Management: An International Journal*, **5(2)**, pp. 182–195, 2010.

[12] Phelps, R., and Hase, S., Complexity and action research: exploring the theoretical and methodological connections. *Educational Action Research*, **10(3)**, pp. 507–524, 2002.

[13] Waterman, H., Tillen, D., Dickson, R. & De Koning, K., Action research: a systematic review and guidance for assessment. *Health Technology Assessment*, **5(23), pp.** 1–166, 2001.

[14] Anderson, G.L. & Herr, K., The new paradigm wars: is there room for rigorous practitioner knowledge in schools and universities? *Educational Researcher*, **28(12)**, pp. 12–40, 1999.

[15] Newton, P. & Burgess, D., Exploring types of educational action research: implications for research validity, *International Journal of Qualitative Methods*, **7(4)**, pp. 18–30, 2008.

[16] Holter, I.M. & Schwartz-Barcott, C., Action research: what is it? how has it been used and how can it be used in nursing? *Journal of Advanced Nursing*, **18(2)**, pp. 298–304, 1993.

The relationship between employee health, quality culture and organizational effectiveness: findings from the literature

R. Sadia
Department of Industrial Engineering and Management, Shenkar College of Engineering and Design, Israel

Abstract

Organizations are complex social systems, and as such, they ought to be approached through a systems thinking viewpoint in order to understand the whole rather than just the parts. The purpose of this literature review is to outline the complex interdependent relationship between organizational effectiveness, employees' health and quality culture. This review will open with the notion of organizations as complex social systems. Then, it will describe possible applications of systems thinking in two parts: The first part will discuss quality and quality culture, and the second part will discuss employees' health in the organization. The review will include the application of system thinking and will show the important contribution of the system approach to the achievement of organizational effectiveness.
Keywords: employee health, organization health and effectiveness, stress, systems thinking.

1 Introduction

Numerous theories and studies deal with various aspects of improving organization's effectiveness as it undergoes change. In many studies ([1–6] and others), the emphasis has been on quality directed towards attaining customer satisfaction as the primary long-term objective above and beyond maximizing shareholder value [3]. Focusing on quality rather than focusing on organization's effectiveness represents a paradigm shift in management thinking. The quality management paradigm is based on behavioral approaches that can enhance performance by recognizing the employee as a customer and emphasizing

WIT Transactions on State of the Art in Science and Engineering, Vol 90, © 2016 WIT Press
www.witpress.com, ISSN 1755-8336 (on-line)
doi:10.2495/978-1-78466-155-7/017

teamwork and participation as vehicles for job satisfaction, motivation, and organizational performance [7–9]. Despite the presumed benefits from this paradigmatic stress on quality, there has been little emphasis on the importance of the employee's quality of life at work as a quality component.

The importance of considering all the components in an organization as parts of a whole system has triggered consideration of the systems approach, a paradigm that views a system as a group of interdependent, interacting parts [10]. The concept of systems thinking has been regarded as an important characteristic of total quality formulations [4, 11]. It was also accepted later as an important core value of performance excellence by the MBNQA (Malcolm Baldrige National Quality Award) [12] and by ISO 9001: 2000 [13]. Despite this, employee health is viewed as a separate issue and not sufficiently important for strategic quality initiatives. Only lately has the interaction between employee health and organizational effectiveness started to intrigue researchers who have begun to turn their attention to ways of advancing employee health in order to improve an organization's productivity [14–17]. The interaction between quality culture and employee health is also a subject of recent interest [18–20]. Unfortunately, the literature generated by the interest in these three constructs, employee health, quality culture and organizational effectiveness, does not focus on the systems approach as a tool for understanding the relationships between these three concepts.

In the following paper, the literature review deals with quality programs, employee health and organizational effectiveness. A discussion of systems thinking and its basic concepts follows. Based on the literature and the use of the systems thinking tools, the key inter-relationships between many variables that relate to quality culture, employee health and organizational effectiveness are presented.

2 Literature review

2.1 Quality programs, quality culture and organizational effectiveness

In the 1990s, much of the literature concerning quality in organizations was devoted to the study of the relationships between quality and other important measures of organizational success, such as productivity [21]; profitability [22]; market value [23]; competitive advantage [24]; and organizational performance [9, 25, 26]. The reason for this focus may have been that the previously prevailing mindset in which quality by itself was not regarded as the end of a process, but as a means to something else. Only later did organizations discover that certain quality approaches, such as zero-defects or six-sigma might also be associated with effectiveness goals [27].

Many researchers investigated the reasons for the lack of success in implementing TQM in the workplace [7, 28–30]. Some [8, 28, 31] concluded that the low rate of TQM success was a result of focusing mostly on the 'hard' issues and neglecting the 'soft' issues when implementing TQM. 'Hard' issues were regarded [8] as core quality practices and had a more technical orientation

[32]. Among these issues were quality information, process management, product design and statistical process control. The 'soft' issues were [8] infrastructure quality practices and involved more of the social and behavioral attributes of quality management [32]. Among these issues were: employer–employee relationships, top management support, customer involvement and other human relationships. At the root of the many cases of failure in the process of TQM implementations was the emphasis on quality products rather than quality interactions, and not viewing the employee as one of the main customers and the most important stakeholder of the organization [33]. Failure in implementing TQM could mostly be related to the elements that support the implementation process like the lack of support of the company leadership, rather than to the quality practices themselves [28].

Most of the literature that deals with organizational quality culture focuses on the need for a paradigm change with respect to the prevailing concepts and attitudes that are required in order for quality programs to work [34, 35]. A different approach to this issue is presented by Detert *et al.* [36] who constructed a general framework in terms of culture that can be linked to improvement initiatives in organizations. In their study, they demonstrated a link between this framework and TQM values and beliefs and presented eight dimensions that are most frequently discussed in the literature, like ideas about stability versus change and others.

2.2 Employee health and organizational effectiveness

At the beginning of the 1970s, organizations moved from viewing workforce health in its relation to organizational performance to a more proactive approach, and designed programs that reinforced lifestyle changes [37, 38]. The main concept underlying these programs was that the individual's lifestyle contributed directly to a person's health and organizations need to help individuals change their unhealthy lifestyles. While these programs continue to flourish, other researchers pointed to the link between environmental factors and employee health, especially the effect that stress (an individual's physical and mental reaction to environmental demands or pressures) has on an employee's health [5, 39, 40].

The literature concerning the relationship between health initiatives and organizational effectiveness is far from being consistent in terms of the methods, terms and approaches that measure and evaluate organizational effectiveness and what influences it. The most frequently used term in the health promotion literature is *productivity* rather than *effectiveness* or *performance*. Even when authors use the word *productivity,* they are actually referring to one measure they believe represents productivity – absenteeism. While acknowledging this shortcoming, they explain it by the fact that methodologies for measuring productivity are lacking.

Later, certain researchers have started to view health in a broader context, specifically when dealing with an organization's health. Grawitch *et al.* [41] identified five general categories of healthy workplace practices in organizations that were linked to employee well-being and organizational improvements:

work-life balance, employee growth and development, health and safety, recognition, and employee involvement. According to Rosen [42], in healthy companies, products and profits are not the immediate goal; they are the result of doing everything else right. Organizational success, improved quality, better service and competitive advantage are the by-products of shared values and collective efforts.

2.3 Systems thinking

The systems approach distinguishes itself from the more traditional analytical approaches by emphasizing the interactions and connections between the different components of a system. The interactions of the parts become more relevant to understanding the system than understanding the parts. According to systems thinking, system behavior results from the effects of complex feedback systems.

2.3.1 System thinking tools

System thinking tools may be regarded as a way of describing a story. The picture shows the elements that depict the fundamental issues of a situation with arrows between them representing the influence that one element has on another. These links reveal cycles that continually repeat themselves, causing the situations to either get better or worse over time. These loops comprise a representation of causality called a feedback loop [43]. An arrow between two variables might describe a positive effect, meaning that if one variable increases (decreases) so does the affected variable, all else being equal. A negative effect is caused when one variable increases (decreases) and causes the other variable to decrease (increase), all else being equal.

There are two sets of behavioral structures: the reinforcing feedback loop and the balancing feedback loop. Figure 1 shows a reinforcing process, which generates exponential growth or collapse at an ever-increasing rate (Fig. 1):

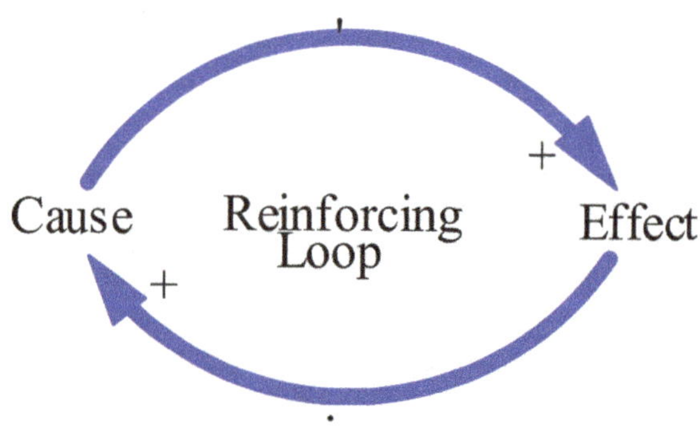

Figure 1: Reinforcing feedback loop.

The reinforcing feedback loop in Fig. 1 demonstrates how cause and effect reinforce each other over time, such that an increase in the cause brings about an increase in the effect, and vice versa.

Balancing processes generate forces of resistance that eventually limit growth. All balancing processes contain a self-correcting or governing function that attempts to attain some goal or target. Figure 2 shows how balancing processes are characterized by a gap between actual and desired performance.

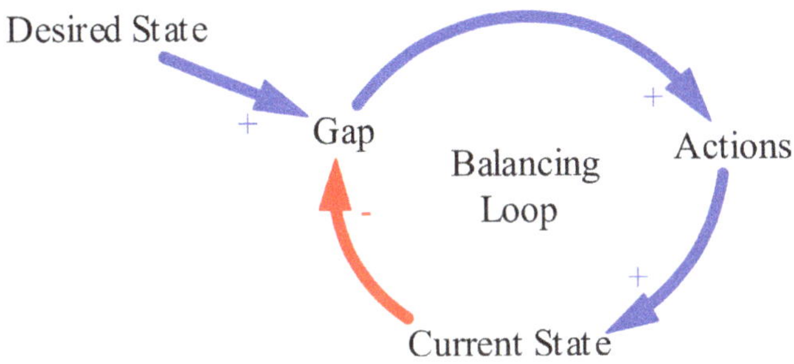

Figure 2: Balancing feedback loop.

As it can be seen in Fig. 2, when the gap between the desired and the current state is large, more actions must be taken to make a large improvement in the current state. A large improvement in the current state, then, lowers the gap, thus requiring fewer actions for the current state.

It is also important to notice that there may be delays, possibly of different durations, associated with the interactions between the parts of a system. These delays will determine the length of time it takes for the characteristics of the structure to become evident.

To understand the complexity of a problem, the problem has to be identified by describing verbally the relationships between all the components of the system and then the conceptual model is built, describing visually these relationships using casual feedback loops and stating the dynamics hypotheses that describe the behavior of the system over time.

3 The conceptual model

The aim of this literature review was to build a holistic framework that will focus simultaneously on promoting organizational and employees' outcomes, assuming that a balance between these two outcomes is the best for both the organization and its employees in the long-run. The literature review provides the researchers with a tool to explore ways in which quality culture and employee health interact and mutually influence each other and then impact organizational effectiveness. The estimated hypothesis is worded using dynamic hypothesis. A dynamic hypothesis is a potential explanation of how the system structure causes the observed dynamic behavior as a result of the interactions between several factors in the system over time [44]. The following dynamic hypotheses were developed following the literature review summarized previously. The relevant studies that provided inspirations for each of the dynamic hypotheses will be mentioned at the end of each description.

3.1 The dynamic hypotheses

The first dynamic hypothesis (Learning Loop in Fig. 3) says that to operate effectively in an organization, a manager or employee should be involved in a learning process that influences his mental models (perceptions), so that he can develop new ways of thinking which can lead to a greater acceptance of change and growth. Assuming that there is a desire for this kind of learning and openness, this may result in closing the gap between the current mental models (defined by the collective skills, knowledge and experience) one has and the new mental models one encounters. The effort to close this gap will lead to decisions and then to actions that one needs to take in order to achieve more experience, skills and knowledge. This effort in turn leads to more learning and a greater degree of openness leading to new skills and the motivation to minimize the gap between the desired and achieved level of learning and openness (based on [43, 45]).

The second dynamic hypothesis (Stress/Control Loop in Fig. 3) describes that the more experience gained through the development of one's skills and knowledge, the more control one has over his life. The greater the alignment between the sense of control (represented by experience, skills and knowledge) and the desire for control (which is measured in the same way), the less the - perceived gap between the desired control and the perceived control. Consequently, less stress is introduced into one's life. This in turn encourages a person to look for more experience and more learning opportunities (based on [5, 14, 39, 40]).

The third dynamic hypothesis (Job Satisfaction/Health Loop in Fig. 3) phrases that the more experience, skills and knowledge the individual gains, the greater the job satisfaction one feels as long as the perception of the level of collaboration, the level of decision latitude, and the level of participation and involvement are close to the levels that one desires. According to the literature, these concepts define important components of job satisfaction. The closer the level of job satisfaction is to the desired level of job satisfaction, the more one feels pleased (this relationship is assumed as part of the link between job satisfaction gap and employee health). Subsequently, this pleasure positively influences one's state of health. If an employee experiences an increased sense of well-being, he tends to be less absent and more productive at work. This in turn provides him with more opportunities to take more actions and develop more skills and knowledge (based on [5, 16, 17, 39]).

The fourth dynamic hypothesis (Stress Recovery Loop in Fig. 3) says that lack of knowledge and skills leading to a feeling of lack of control increases stress up. If this build-up is accompanied by a low perceived level of decision latitude, then stress accumulates even more. This causes a person to experience a greater distance from his/her desired comfort zone. The greater the perceived gap with respect to one's comfort zone, the greater the chances that he will get sick, a state that provides him with an escape route to lower his/her stress (based on [5, 39, 40]). This dynamic hypothesis deals with the linkage between stress and illness.

The fifth dynamic hypothesis (Absenteeism/Stress Loop in Fig. 3) explains that the low health status of an employee may lead to absenteeism reducing a person's productivity and the ability to acquire more skills, experience and knowledge, which increases his/her attendance gap, subsequently causing the person to be less productive and therefore, hindering one's ability to acquire more skills, experience and knowledge. This state, in turn, increases the perceived control gap, leading to even greater stress. As stress accumulates, the perceived comfort zone gap increases, eventually causing even more illness (based on [5, 16, 17, 39]).

The sixth dynamic hypothesis (Motivation Loop in Fig. 3) says that the greater the job satisfaction one experiences, the more motivation one has at work. When motivation is high, commitment is also high, which positively influences the actual job performance. This holds true as long as the level of communication needs with the other workers is commensurate with the job requirements. If actual job performance is high, the job quality level is high, resulting in a higher level of actual organizational effectiveness. When the actual organizational effectiveness rises, the effectiveness gap decreases and management satisfaction increases. This positively influences employee satisfaction and motivation (based on [20]).

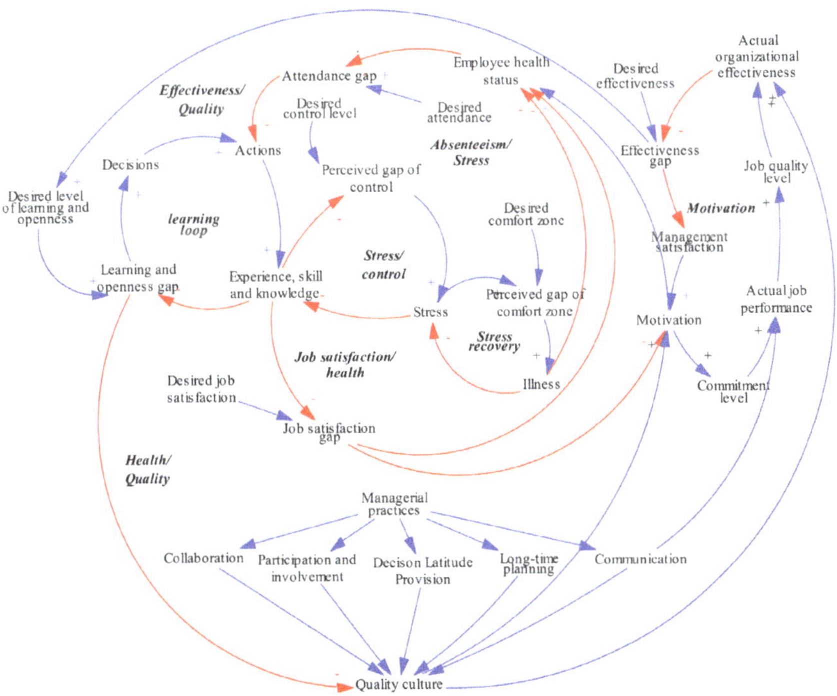

Figure 3: The full qualitative model derived from the literature.

The seventh dynamic hypothesis (Health/Quality Loop and Effectiveness/ Quality Loop in Fig. 3) describes that quality culture increases with an increase in the implementation of such managerial practices as collaboration, expanded decision latitude, participation and involvement, better communication and long-term planning. Quality culture positively impacts employee motivation as well employee health status and actual organizational effectiveness. These developments decrease employee attendance gaps (the gap between the desired attendance at work and the actual attendance) and the organizational effectiveness gap. This in turn increases the actions at work taken by the employee and increases his current experience, skills and knowledge. More experience and knowledge decrease the openness gap (the gap between the current experience, skills and knowledge and the desired level of learning and openness) and finally increases quality culture (based on [20, 34, 36, 47]).

3.2 The full qualitative model

Based on the dynamic hypotheses that were extracted from the literature, a conceptual model was constructed, demonstrating the linkage among the variables and processes that are significant components of quality culture, employee health and organizational effectiveness. Figure 2 presents the unified model, which was constructed by linking all the previously described loops.

While all the relationships and connections in Fig. 3 are important for an organization, one can conclude that the most significant loops for the purpose of this research are the Health/Quality and the Effectiveness/Quality loops and the Motivation loop, since they highlight the connection between quality culture employee health and organizational effectiveness and how they mutually influence each other.

4 Conclusions and recommendations

This paper contributes to the complex systems literature by exploring the relationships between several components that are usually explored individually in research. The review starts from the notion that organizations are social systems. Management practices, organizational culture and organizational values are not separate entities that can be looked in isolation and measured in relation to the organizational and employees outcome, nor can the organizational outcome be detached from employees' outcome, since they are deeply interrelated. In complex systems, especially in complex social systems like organizations, systems thinking can provide a better and more useful approach to capture the dynamic of the system.

Viewing organizations as systems requires the use of interdisciplinary tools that are aimed at improving organizational and employee outcomes. Neither health nor effectiveness is the objective construct. The way they are perceived is always related to their cultural context. What can be considered effective in a certain organization, in a certain situation or environment, could easily be considered ineffective in a different situation. The same holds for health. When we consider any social system one have to take into account psychological and social measures, beside the objective, quantitative measures which are commonly used.

Both health and effectiveness are multi-dimensional phenomena involving interdependent physical, psychological and social aspects. Each one of these aspects can affect the individual health or the organizational effectiveness, or be affected by them. These multiple dimensions of health and effectiveness will affect one another and will bring the system to its optimal state only when they are balanced and integrated.

The use of feedback loops in building the conceptual model offers the ability to identify ongoing patterns in organizations according to the literature. On the basis of this review, case studies in various organizations with different kinds of cultures and backgrounds should be studied and investigated to determine whether the relations described in the conceptual model exist in real-life social systems. Other studies can go even further and translate the conceptual model into a formal, quantitative representation. After this quantification, it can be simulated so that one can understand its behavior over time. Overall, using a system approach for constructing a conceptual model which depicts the complexity of the organization can prove useful for understanding the linkage between organizational effectiveness, employee health and quality culture.

References

[1] Jiang, J.Y. & Liu, C.W., High performance work systems and organizational effectiveness: The mediating role of social capital. *Human Resource Management Review*, **25**, pp. 126–137, 2015.

[2] Matthews, J.R., Assessing organizational effectiveness: the role of performance measures. *Library Quarterly*, **81(1)**, pp. 83–110, 2011.

[3] Koys, J.D., The effects of employee satisfaction, organizational citizenship behavior, and turnover on organizational effectiveness: a unit-level, longitudinal study. *Personnel Psychology*, **54**, pp. 101–114, 2001.

[4] Deming, W.E., *The New Economy*, MIT Center for Advanced Engineering Study: Cambridge, MA, 1994.

[5] Karasek, R. &Theorell, T., *Healthy Work: Stress, Productivity, and the Reconstruction of Working Life*. Basic Books, Inc: New York, NY, 1990.

[6] Argyris, C., *Overcoming Organizational Defenses. Facilitating Organizational Learning*, Allyn and Bacon: Boston, MA, 1990.

[7] Beer, M., Why total quality management programs do not persist: the role of management quality and implications for leading a TQM transformation. *Decision Sciences*, **34(4)**, pp. 623–641, 2003.

[8] Fotopoulos, B.C. & Psomas, L.E., The impact of "soft" and "hard" TQM elements on quality management results. *International Journal of Quality & Reliability Management*, **26(2)**, pp. 150–163, 2009.

[9] Anderson, M. & Sohal, A.S., A study of the relationship between quality management practices and performance in small businesses. *International Journal of Quality and Reliability Management*, **16(9)**, pp. 859–877, 1999.

[10] Sterman, D.J., *Business Dynamics, Systems Thinking and Modeling for a Complex World,* Irwin McGraw-Hill: New York, NY, 2000.

[11] Deming, W.E., *Out of the Crisis*, MIT Center for Advanced Engineering Study: Cambridge, MA, 1986.

[12] Malcolm Baldrige National Quality Award (MBNQA), *Criteria for Performance Excellence*, NIST: Washington, DC, 2008.

[13] International Organization for Standardization, *Quality Management Systems – Requirements, ISO 9001:2000,* ISO Publication: Switzerland, 2000.

[14] Hasson, H., Brisson, C., Guérin, S., Gilbert-Ouimet, M., Bari-Gingras, G., Vézina, M. & Bourbonnais, R., An organizational-level occupational health intervention: employee perceptions of exposure to changes, and psychosocial outcomes. *Work & Stress*, **28(2)**, pp. 179–197, 2014.

[15] Wilson, G.M., Basta, B.T., Bynum, H.B., DeJoy, M.D., Vandenberg, J.R. & Dishman, K.R., Do intervention fidelity and dose influence outcomes? Results from the move to improve worksite physical activity program. *Health Education Research*, **25(2)**, pp. 294–305, 2010.

[16] Riedel, E.J., Lynch, W., Baase, C., Hymel, P. & Peterson, W.K., The effect of disease prevention and health promotion on workplace productivity: a literature review. *American Journal of Health Promotion,* **15(3)**, pp. 167–191, 2001.

[17] Berger, L.M., Murray, F.J., Xu, J. & Pauly, M., Alternative valuations of work loss and productivity. *Journal of Occupational and Environmental Medicine*, **43(1)**, pp. 18–24, 2001.

[18] Lagrosen, Y., Bäckström, I. & Lagrosen, S., The relationship between quality management and health-exploring the underlying dimensions. *International Journal of Productivity and Quality Management*, **5(2)**, pp. 109–123, 2010.

[19] Baptiste, N.R., Tightening the link between employee well-being at work and performance: a new dimension for HRM. *Management Decision,* **46(2)**, pp. 284–309, 2008.

[20] Browne, H.J., Benchmarking HRM practices in healthy work organizations. *American Business Review,* **18(2)**, pp. 54–61, 2000.

[21] Benavides-Chicon, C.G. & Ortega, B., The impact of quality management on productivity in the hospitality sector. *International Journal of Hospitality Management*, **42**, pp. 165–173, 2014.

[22] Singhal, R.V. & Hendricks, B.K., The financial justification of TQM. *Center for Quality of Management Journal*, **8(1)**, pp. 1–19, 1999.

[23] Lin, C. & Su, C., The Taiwan nation quality award and market value of the firms: an empirical study. *Production Economics*, **144**, pp. 57–67, 2013.

[24] Powell, C.T. Total quality management as competitive advantage: a review and empirical study. *Strategic Management Journal*, **16**, pp. 15–37, 1995.

[25] Macinati, S.M., The relationship between quality management systems and organizational performance in the Italian National Health Service. *Health Policy*, **85**, pp. 228–241, 2008.

[26] Samson, D. & Terziovski, M., The relationship between total quality management practices and operational performance. *Journal of Operations Management*, **17**, pp. 393–409, 1999.

[27] Sudit, F.E., *Effectiveness, Quality and Efficiency: A Management Oriented Approach*, Kluwer Academic Publishers: Boston, MA, 1996.

[28] Yamada, T.T., Poltronieri, F.C., Gambi, N. do L. & Gerolamo, C.M., Why does the implementation of quality management practices fail? A qualitative study of barriers in Brazilian companies. *Social and Behavioral Sciences*, **81**, pp. 366–370, 2013.

[29] Page, R. & Curry A., TQM – a holistic view. *The TQM Magazine,* **12(1)**, pp. 11–17, 2000.

[30] Saad, G.H. & Siha, S., Managing quality: critical links and a contingency model. *International Journal of Operations & Production Management*, **20(10)**, pp. 1146–1163, 2000.

[31] Savolainen, T., Towards a new workplace culture: development strategies for employer–employee relations. *Journal of Workplace Learning,* **12(8)**, pp. 318–326, 2000.

[32] Naor, M., Goldstein M.S., Linderman, W.K. & Schroeder, G.R. The role of culture as driver of quality management and performance: infrastructure versus core quality practices. *Decision Sciences*, **39(4)**, pp. 671–702, 2008.

[33] Sadia, R., *A System Dynamics Approach Linking Employee Health, Quality Culture and Organizational Effectiveness*, PhD Dissertation, Virginia Polytechnic Institute and State University, Virginia, 2006.

[34] Harvey, L. & Stensaker, B., Quality culture: understanding, boundaries and linkages. *European Journal of Education*, **43(4)**, pp. 427–442, 2008. doi: http://dx.doi.org/10.1111/j.1465-3435.2008.00367.x

[35] Dellana, A.S. & Hauser, D.R., Towards defining the quality culture. *Engineering Management Journal*, **11(2)**, pp. 11–15, 1999.

[36] Detert, R.J., Schroeder, G.R. & Mauriel, J.J., A framework for linking culture and improvement initiatives in organizations. *Academy of Management Review*, **25(4)**, pp. 850–853, 2000.

[37] Goetzel, Z.R., The financial impact of health promotion and disease prevention programs – why is it so hard to prove value? *American Journal of Health Promotion,* **15(5)**, pp. 277–280, 2001.

[38] Aldana, G.S., Financial impact of health promotion programs: A comprehensive review of the literature. *American Journal of Health Promotion*, **15(5)** pp. 296–320, 2001.

[39] Abdullah, D.N.M.A. & Lee, O., Effects of wellness programs on job satisfaction, stress and absenteeism between two groups of employees (attended and not attended). *Social and Behavioral Sciences*, **65**, pp. 479–484, 2012.

[40] Anderzen, I. & Arnetz, B.B., The impact of a prospective survey-based workplace intervention program on employee health, biologic stress markers, and organizational productivity. *Journal of Occupational Environmental Medicine*, **47(7)**, pp. 671–682, 2005.

[41] Grawitch, M.J., Gottschalk, M. & Munz, D.C., The path to a healthy workplace: a critical review linking healthy workplace practices, employee well-being, and organizational improvements. *Consulting Psychology Journal: Practice and Research*, **58(3)**, pp. 129–147, 2006.

[42] Rosen, H.R., *The Healthy Company*, Jeremy P. Tarcher/Perigee: New York, NY, 1992.

[43] Senge, M.P., *The Fifth Discipline*, Currency Doubleday: New York, NY, 1990.

[44] Oliva, R., Empirical validation of a dynamic hypothesis. *Proc. of the Int. Conf. of System Dynamics*, eds. G.P. Richardson & Sterman, J. D, System Dynamics Society: Cambridge, MA, pp. 405–408, 1996.

[45] Forrester, W.J., System dynamics and the lessons of 35 years. In *A Systems-Based approach to policy making,* ed. K.B. De Greens, Kluwer Academic Publishers: Cambridge, MA, 1993.

Systems engineering beliefs: contemplating personal perceptions regarding state of the art

B. E. White
Complexity Are Us ß Systems Engineering Strategies, USA

Abstract

Systems engineering (SE) theories and practices for effectively dealing with complex systems are still evolving. Principal stakeholders are integral to complex systems, so we cannot predict or prespecify how the system changes. Every systems engineer has his own professional viewpoint derived from cultural background, upbringing/education and experience/maturation. The author has decided how all this fits his career and current 'mindsight'. He encourages other systems' engineers to examine and share their thoughts to help attain a more collective and forward looking view of systems engineering to advance the field. This paper provides a framework for system engineers to contemplate whether they are of the traditional system of systems; enterprise; or the complex type. Perspectives on SE attributes are offered to assist any systems engineer in characterizing his belief system, stylistic approach, and overall effectiveness in engineering complex systems. Systems engineers might reflect on what applies to them and to what relative extent. If a particular set of attributes dominates, is that appropriate? What changes in outlook, style, and emphasis might make a more effective systems engineer? Several existence categories are suggested and proportioned through each of life's stages. System engineers may contemplate how these trends have affected their past behaviors, and how possible re-directions might improve their future accomplishments. Thinking through what we have done, are doing, and want to do as systems' engineers is recommended. Anyone's chosen way is no better than any other possibility. It depends on a multitude of circumstances in each of our lives.
Keywords: belief system, complex system, enterprise, religion, system of systems, systems engineering.

www.witpress.com, ISSN 1755-8336 (on-line)
doi:10.2495/978-1-78466-155-7/018

1 Introduction

Many systems' engineers practice the art of systems engineering (SE) following their own belief system with respect to the spectrum of SE methods of which they are aware or might contemplate applying. Since this is analogous to one's religious beliefs, some parallels are drawn between religion's belief systems, viz., Theism, Agnosticism, and Atheism, and (classical, conventional, or traditional) SE, System of Systems (SoS) Engineering (SoSE), Enterprise SE (ESE), and Complex Systems Engineering (CSE) [1]. A system engineer's existential categories, viz., authorities, exposure, learning, paths, and contributions, are also discussed as lifetime threads that influence SE.

It is extremely difficult, if not impossible, to provide theoretical or experimental evidence that the opinions expressed herein are valid. Instead, without any so-called proofs, a dialog involving fictitious characters is used to provide balanced views. Most readers will find this interesting and somewhat entertaining. The principal goal is to stimulate reflection and creative thinking about improving SE approaches to our most difficult problems. Unfortunately, too many systems' engineers set their sights much lower and are content to immerse themselves in more limited situations.

Table 1 summarizes the cast of characters. Each statement will be preceded by the maker's character name to better flag the point of view espoused.

2 Systems engineering attributes

Facilitator: Thank you for agreeing to meet. We can all benefit from discussing our belief systems and from where they derived in improving our SE practice. The profession needs to better handle not only your individual and organization's needs but also the world's more pressing problems. Let's first focus on religious belief systems. Other topics will naturally follow. Who will begin the discussion?

2.1 Religious belief systems

Theist: All right, I'll start. Theism is the most common religious belief system. According to the Encarta Dictionary: theism has two meanings: (1) Belief in God, i.e. belief that one God created and rules humans in the world and (2) belief in God or Gods. Theists believe God exists and have, in effect, adopted pantheism: (1) belief that God is everything and (2) belief in all deities. I characterize the religious philosophy of theists and pantheists by *tradition*, *acceptance*, *devotion*, *faith*, *dogma*, and *ritual*, and strongly support these tenets.

Agnostic: Agnosticism is much less common. Encarta says agnostics view God's existence as unprovable. Agnostics are neutral about God's existence and have effectively adopted religious naturalism [2]: (1) Relating to (belief in, teaching or practice of) religion and (2) believing in a higher being. Religious naturalism seems to fall between agnosticism and theism. I characterize

Table 1: Cast of characters in dialog.

Character name	Character type	Role in dialogue
Agnostic	Ambivalent about God	Humble in not being sure what to believe
Atheist	Does not believe in gods of any kind	Favors rationale thought; beliefs must make sense logically
Boss	Autocratic but loyal to and defender from outsiders of people in organization	Advocate of hierarchical organizational structure and pleasing those at the top
Coach	Believer in positively shaping personal character through competitive sports	Shows analogies between lessons of life and good systems engineering
Complex Systems Engineer	Focused on the most general form of systems engineering and advising on techniques that might be more effective	Emphasizes complex systems behaviors and complex systems engineering principles
Conservative Advocate	Believes in individual freedom, right to life, and limited government	Straight-forward, into discipline, and proud of being a Christian
Enterprise Systems Engineer	Takes broad viewpoint embracing homeostatic systems not necessarily containing only component systems	Eager to learn more about what complex systems engineering techniques might be more effective
Facilitator	Concerned with fundamental issues, abstract ideas, and significant progress toward solving human made problems	Leads in trying to stimulate dialogue that may help systems engineers appreciate how belief systems shape approaches to the profession
Follower	Reads lots of advanced systems engineering papers and asks for more	Encourages further pursuit of effective complex systems engineering methods
Liberal Supporter	Concerned with human welfare and favoring government action	Voice of compassion
Lover	Attracted to kind, well-rounded, hardworking intellectuals who can also play	Reminds friends that a balanced life is more important than workaholic pursuits

Table 1: Continued.

Character name	Character type	Role in dialogue
Parent	Loves and cares for their kids but leaves a lot up to school teachers	Defensive, recognizing only limited responsibility for rearing children
School Teacher	Remembered as favorite of many students even much later in life	Endures as tremendous influence in attitudes about life and career development
System of Systems Engineer	Understands increased difficulty in engineering complicated and more complex systems containing component systems	Largely oriented toward military systems of systems characteristic of principal U.S. Department of Defense concerns
Systems Engineer	Classical, conventional, or traditional systems engineering practitioner	Defends conventional systems engineering methods and is very reluctant to change his/her prevailing mindset
Systems Thinker	Knowledgeable about complex systems, enterprises, systems of systems, and systems; smart, and dispassionate	Provider of balanced and objective viewpoints and ideas for achieving deeper research and more effective practices
Theist	Believes in God	Accepts religious beliefs primarily on faith

agnostics by *humility*, *ambivalence*, *uncertainty*, *openness*, *learning*, and *analysis*.

Atheist: Atheism is the rarest. Atheists do not believe in a personal God or any gods at all. We are characterized by *rationality*, *skepticism*, *humanism*, *experimentation*, *study*, and *synthesis*. In my experience, most people feel that atheism is not a religious belief. However, in the sense of being 'a belief or practice forming part of someone's thought about ¼ a divine being' (Oxford American College Dictionary), it really is!

Conservative advocate: Christianity is the true religion, as evidenced by Jesus's good works, his resurrection, and the profound impact belief in Jesus Christ has had on the world. I support the view of theists.

Liberal Supporter: Being a Unitarian Universalist, all three views have merit. Why can't we respect anyone's belief system? The world should share beliefs openly, subjugate harsh judgments, and embrace uplifting ideas.

Systems Thinker: This is a reasonable beginning, and I commend the Facilitator for instigating this meeting.

Facilitator: I think so, too, and thank you. Next let's pose typical attributes associated with different kinds of SE and relate those attributes to belief systems.

2.2 SE attributes

Systems engineer: Attributes associated with theists are *risks*, *reductionism*, *technology*, *requirements*, and *doctrine*. SE is structured, ordered, predictable, and malleable. Complicated problems can be effectively addressed provided sufficient care is taken. Requirements can be clearly, unambiguously, and promptly defined by the user, the customer, and the SE team. Problems can be subdivided, optimizing each piece separately, to be reassembled without perturbing the collective effect. The latest technology is applied to the extent possible. Risks needing mitigation are recognized and handled. A very logical lifecycle process is followed.

Complex systems engineer: Allow me to take issue with a few of your points. In your more complex problems, haven't you noticed how difficult it is for anyone to define unchanging requirements? Reductionism doesn't work because by the time you apply your rigid, linear, lifecycle process, the system and its environment has evolved.

Enterprise systems engineer: Right, a broader approach must be taken, especially in enterprises that are more general than regular systems or more complex than SoS's. I want to learn about more effective SE techniques.

System of systems engineer: SE is just fine! It just needs to be extended to accommodate collections of systems. The U.S. Department of Defense (DoD) recognized this long ago, and mounted extensive efforts to concentrate on the most important area, SoS's and SoSE.

Enterprise systems engineer: SE attributes associated with agnostics are *independence*, *innovation*, *networking*, and *science*. Agnostics view SE as more complex and challenging. There is more independence of thought, innovation in trying new solutions, networking with colleagues to share information, and the application of scientific facts, at least until disproven.

Complex systems engineer: SE attributes associated with atheists are *isolation*, *distrust*, *contemplation*, and *outreach*. Atheists view SE as very complex, much more than just complicated. They tend to: be ostracized, particularly if abrasive; often express minority opinions or resist management; distrust conventional methods of treating difficult problems; continually look for more effective pathways; be thoughtful and examine situations with multiple perspectives; look for fact-based evidence in others' work that might support their own notions or give them ideas; and interact with their receptive and intellectually stimulating colleagues.

System of systems engineer: SE atheists reap what they deserve! Why don't they contribute something useful to the program instead of causing disruption?

Systems engineer: I agree. What we do works well enough. Do not judge us!

Facilitator: Actually, I think we've all wrestled with most of these thoughts, and please, let's concentrate on evaluating and sharing comments on the ideas expressed, and neither insult nor question the integrity of the speaker.

2.3 SoSE attributes

System of systems engineer: SoSE attributes associated with theists are *modularity*, *security*, *hierarchy*, and *hubris*. Theists view SoSE as an extension of SE. Their tendency is to continue trying traditional SE techniques. Here, they run into difficulty because an SoS is a collection of independent systems whose interactions can be much more involved. This leads theists to emphasize modularity, a good organizing principle. Theists are concerned with security to counter lack of trust. Hierarchy takes a larger role because an SoS rides above its component systems, and processes are driven by authorities with a top-down point-of-view. Theists exhibit hubris in their comfort with SE methods.

Enterprise systems engineer: I agree and suggest that SoSE attributes associated with agnostics are POET (Political, Operational, Economic, and Technological [factors]), *discussions*, *heuristics*, and *layering*. There is less hubris compared to theists because agnostics better appreciate the difficulties involved. They share perspectives in understanding SoS's; look for rules-of-thumb to improve decision-making; and bypass modularity to advocate a more effective *layered* architecture.

Complex systems engineer: Very well said. SoSE attributes associated with atheists include *opportunities*, *capabilities*, *interoperability*, and *systems thinking*. Atheists know that opportunities abound and can lead to better solutions. They: are not wedded to requirements but are interested in increasing capabilities; value interoperability, not only within a SoS but also across SoS's; and employ systems thinking, including what has been omitted and what might not even be there – that should be.

Systems thinker: Glad you mentioned my realm. Holistic thinking is critical, not only in your CSE but also in ESE, SoSE, and where it is most lacking, SE.

Systems engineer: I'm not sure what holistic means but I think you're being unfair to our practice. If we take care of the parts, the whole takes care of itself!

Facilitator: Again, let's all be more careful how we treat each other.

2.4 ESE/CSE attributes

Complex systems engineer: ESE/CSE attributes associated with theists include *complication*, *optimization*, *point-to-point*, and *dicta*. Theists: use the word complex when they mean complicated; discount complexity theory; try to optimize a CS instead of balancing competing interests to achieve results that satisfies principal stakeholders; gravitate to point-to-point communications which is not scalable in a large network, there being too many combinations to detail and update; and are not interested in studying complexity, CSs, or ESE/CSE techniques but readily embrace dicta on what higher authorities think works.

Systems engineer: Didn't you hear the Facilitator just now? Your presentation style is brutal! Nonetheless, I think I understand most of what you say and will consider your remarks carefully, although maybe later with my own team.

Enterprise systems engineer: ESE/CSE attributes associated with agnosticism include *humility*, *balance*, *interactions*, and *simplicity*. Agnostics: view

ESE/CSE with respect; bring humility to the study and contemplation of, and attempted interventions in, a CS; are more inclined to try to achieve balance among competing thrusts; recognize the importance of understanding interactions among CS portions that strongly affect overall behavior; and try to penetrate through complexity to reach simplicity. Perhaps we need more agnosticism in all of us!

Complex systems engineer: ESE/CSE attributes associated with atheists include *holism*, *trans-disciplines*, *trust*, and *self-organization*. Atheists: view ESE/CSE with pleasure and try to find rational, workable approaches; concentrate on holism; consider POET and transdisciplinary (e.g. cultural, social, psychological, organizational, environmental, political, legal, moral, operational, economic, legacy, etc.) techniques; win the trust of their constituents through intellectual honesty and objectivity (Unfortunately, this is not usually the case in the secular world!); and trust the results of their efforts, at least until they learn more. Self-organization is one of the principal hallmarks of atheists; they believe that collaboration (cooperation and competition) among all protagonists leads to success.

System of systems engineer: Your description is quite self-congratulatory! Actually, I think we operate in much the same way while practicing SoSE, although perhaps with somewhat less emphasis on those attributes you apply.

Facilitator: Although we're on dangerous ground again, I appreciate the somewhat more polite tone. For a moment let's contemplate how far we've come.

2.5 Summary of perspectives achieved

Systems thinker: How well does Table 2 summarize your opinions? SE; SoSE; or ESE/CSE, are labels of the right-most three columns. Theism, Agnosticism, and Atheism are labels of the three rows; the left-most column provides rough definitions of each. The 3×3 matrix formed by the intersection of the rows and right-most columns contains the exponent-coded characterizations of the attributes of the intersecting methods and belief systems.

Complex systems engineer: I couldn't agree more; hope you all agree, too.

Enterprise systems engineer: This is fine with me.

System of systems engineer: I don't know. With more time and effort, I could revise this table considerably, which I may endeavor to do eventually. However, it also seems, to some extent, we're making much ado about nothing.

Systems engineer: Amen. Let's stick to fundamentals and get on with the actual work improving our systems!

Theist: The bottom row this table seems quite reasonable. I've never thought of attributes in such terms.

Agnostic: The middle row is debatable but I'm not yet sure of specific improvements; I'll give that further thought.

Atheist: I have no rational basis for quibbling with the upper row. I'm just happy to be included in this discussion!

Facilitator: So be it for now. Let's now press on to the next topic.

Table 2: Perspectives on systems engineering attributes.

Postulated Belief System	Atheism No personal God No gods at all	Rationality Scepticism Humanism Experimentation Study Synthesis	Isolation^4 Distrust^4 Contemplation^2 Outreach^2	Opportunities^1 Capabilities^2 Interoperability^2 Systems Thinking^2	Holism^1 Trans-Disciplines^1 Trust^1 Self-Organization^1
	Agnosticism There may be a God Religious Naturalism	Humility Ambivalence Uncertainty Openness Learning Analysis	Independence^2 Innovation^2 Networking^2 Science^2	POET^1 Discussions^1 Heuristics^1 Layering^1	Humility^1 Balance^1 Interactions^1 Simplicity^1
	Theism God exists Pantheism Faith	Tradition Acceptance Devotion Faith Dogma Ritual	Risks^4 Reductionism^4 Technology^4 Requirements^4 Doctrine^4	Modularity^3 Security^3 Hierarchy^3 Hubris^4	Complication^4 Optimization^4 Point-to-Point^4 Dicta^4
			SE	SoSE	ESE/CSE
			Systems Engineering Method		

Glossary
SE: Classical, Conventional, or Traditional Systems Engineering
ESE: Enterprise Systems Engineering
SoSE: System of Systems Engineering
CSE: Complex Systems Engineering
POET: Political, Operational, Economic, and Technological [factors]

Exponent Code
1: CSE Principles
2: Laudable Qualities
3: Limitations
4: Things to Mitigate

Note from *Systems Thinker*: I applied a code located at the lower right, where an integer exponent is assigned to each attribute, while considering 13 CSE principles from White [3].

3 Lifetime history of a systems engineer

Facilitator: I have taken some time and effort to think about key categories of existence that have significant influence in the life of a systems engineer. Bear with me, if you will, and let's consider how these threads might be proportioned.

3.1 Category definitions and viewpoints of category representatives

Facilitator: I populated five example categories with instance threads:

- *Authorities*: parents, bosses, project/program leaders/managers, and executives/officials; in business/commerce/industry, and services/ government.
- *Exposure*: family, religion, ethnicity, sports, town/state/country, geography, colleagues, people, nature, travel, events, and politics.
- Learning: elementary/high school, college/university, career, and lifelong.
- Paths: recreation, independent, entertainment, hobbies, relationships, career, investment, and retirement.
- Contributions: accomplishments, artifacts, products, influence, publications, recognition, plaudits, and awards.

I invited representatives of these categories: Boss, Coach, School Teacher, Lover, and Follower, respectively, to join us.

Boss: The dialogue seems academic to me, frankly. What matters is that we get the job done! Someone needs to be in charge/control to direct the effort.

Coach: We need to pay attention to what is important in life as well as work.

School teacher: Lessons taught in school are more enduring than those of the workplace. It comes down to individual effort and following laudable principles.

Lover: Don't miss the fun in life, as the Coach inferred. But all play and no work also makes a dull companion.

Follower: I am intrigued if we are moving toward defining innovative SE techniques. Please remember to share advances and pertinent references with me.

Facilitator: Thanks for your introductory statements. I look forward to more comments. Now let's see how these categories expand/contract as we live.

3.2 Interpretation of proportions in category threads

Facilitator: Everyone's lifetime category pattern is unique. Generally, parents dominate early in life, bosses are influential in early career, but gradually the role of authorities diminishes. World exposure and learning start with birth and continues to increase. The number of path choices starts at zero, increases with age, and contracts near end of life. Similarly, value-added contributions start at zero, gradually increase, accelerate during midcareer, and hopefully continue.

School teacher: Everything is based on learning. It's too bad teachers get paid so little when we have so much influence over our future generations.

Conservative advocate: Private/charter schools are the answer, not more government spending on public education.

Liberal supporter: No, private measures are insufficient to establish worldwide leadership in education and technology.

Coach: Exposure to challenges and opportunities, and learning how to mitigate risks through hard-nosed experiences is vital for healthy growth.

Lover: True happiness is dominated by the quality of one's personal relationships.

Follower: Professional relationships count for those without personal companions.

Systems thinker: Figure 1 illustrates a lifetime (100 years, optimistically) depiction. The vertical bar represents a snapshot of the relative proportions at an arbitrary point (assumed to be between ages 25 and 50). What do you think?

Complex systems engineer: It's uncanny; this could be my own personal profile.

Systems engineer: Roughly speaking, I would interchange the Learning and Exposure threads. In addition, my Paths thread would be significantly narrower.

Facilitator: Okay. Others of you may need more time to assess this. But to save some time for this meeting's end, let's now move to cultural background.

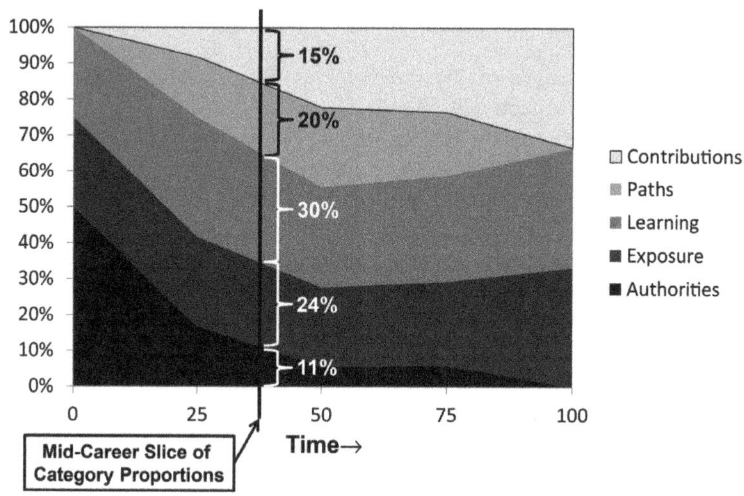

Figure 1: Lifetime proportioning of a systems engineer's existential categories.

4 Cultural background

Facilitator: A person's cultural background as shaped by: parents, family, and ethnicity; and religious, occupational, hometown, state, and country environments. Let's next consider the suggestions of how and why all these early factors influence the later life, behavior, and performance a systems engineer.

4.1 Parents, family, and ethnicity

Complex systems engineer: Babies and very young children understand complexity, albeit at a rudimentary level. We are born into a world where we have no control and can only explore to survive. One soon recognizes their mother and is expected to speak a language. This process is a profound testament to the power of the mind.

Enterprise systems engineer: I certainly agree with that. To what extent do children retain their facility for grasping and coping with complexities as they get older, or is this innate ability discouraged in school so that it eventually atrophies?

Liberal supporter: Parents have a tremendous influence in shaping their children's lifelong personalities, from ages as young as five, instilling behavior, belief systems, and mores. The extended family also has significant impact.

Conservative advocate: The ethnicity of a child's cultural background is also an important factor, since like peoples tend to want to live together. Fortunately, in much of US, Canada, and Europe heterogeneous cultures mitigates this influence.

4.2 Religious, occupational, hometown, state, and country environments

Theist: Religions and occupations practiced by parents are important influences.

Agnostic: A young child cannot make an informed decision about religion, and typically merely accepts what their parents teach. S/he also develops an affinity for mother or father's occupation and wants to become like them.

Atheist: A child begins life as an atheist and quickly becomes a theist due to parental, family, ethnic, and religious environments.

Coach: Most people develop a sense of loyalty to their hometown, their state, and particularly their country. Everyone wants to feel part of the best group.

Follower: One later develops divided loyalties and aligns with other groups. For example, after attending college, starting a career, or getting married.

Parent: This is part of maturing and becoming more independently minded. Many eventually shed much of what they're told but parenting never ceases!

5 Upbringing/education

Facilitator: A child greatly depends on others' guidance. One's response to this influence is critical in shaping further development.

5.1 Fitting into the family mold and teaching right from wrong

School teacher: Parents must teach their children to behave properly. Clearly, such behavior can take many forms based on cultural background.

Parent: Tell me about it! Your job is to motivate kids to learn and to enforce parental instilled discipline while they're in school.

Conservative advocate: If a child does not behave, incentives or disincentives, e.g. a larger allowance or physical punishment, may be applied. Although child-abusive beratements, spankings, or worse, is less prevalent, these practices still exist.

School teacher: Pardon me, again, but another job of parents is to teach children to recognize right from wrong, and to follow a moral path.

Parent: This task we might share and work on together.

Theist: People form their moral outlook, in part, by observing others. The frailties, mistakes, and even immoral actions of someone you like may be overlooked. Similarly, someone you dislike will rarely be emulated.

Conservative advocate: Everyone is dishonest to a degree but, I hope, only in ways that cause no harm. Model the continual pursuit of effective practices.

5.2 Handling problems and growing socially

School teacher: The most important job of parents is to expose their children to increasing levels of difficulty, as we do in their schooling. Unfortunately, some parents overprotect their children, and this can make later life much more difficult.

Parent: Present parties exempted, of course. We might have a parent–teacher meeting on this and other issues in an attempt to see more eye-to-eye.

Complex systems engineer: Children should be taught to understand the problem [4] and ask for advice. One must collaborate in handling difficulties instead of responding in a cavalier, disrespectful, or ineffective manner. Everyone should show appreciation for others' well-intentioned inputs, suggestions, and wisdom.

Parent: Good parents want their children to behave properly in social contexts.

Parent: Sometimes it's important that children be seen and not heard. But a loving absent-child-to-relative connection is most welcome. Having frequent family gatherings, especially on holidays, is helpful in guiding children socially.

School teacher: Children learn social behavior with friends. Hopefully, they've already learned the Golden Rule at home while discovering right and wrong.

Parent: There you go again!

Coach: Competitive sports also instill principles of good teamwork.

Liberal supporter: Every person should lead a productive and happy life, making valuable contributions to society that helps others. Fair and worthwhile competition is fine as long as it is unselfish.

Lover: Bring humility to problems and treat everyone with respect.

Facilitator: Thank you, all. Now let's focus a little more education.

5.3 Elementary/secondary/high school

Parent: What happens to children in Kindergarten through 12th grade (K-12)? The US is falling behind other countries in education quality. We must stop touting our belief: The US is best in everything! What might we do to improve this situation?

Facilitator: Some of us might volunteer in elementary, secondary, and/or high schools to work with teachers and kids to help instill an appreciation of, and hopefully a love for, STEM (Science, Technology, Engineering, and Mathematics) to bolster college majors in engineering.

Complex systems engineer: In many US elementary schools teachers ask a question and expect kids to provide the one right answer. Consequently, children are desensitized to complexity and are rewarded for oversimplified replies.

School teacher: There is a trend in flipping the classroom, where children are expected to pose the questions based on understanding and curiosity consistent with their original complexity view. This helps preparation for an engineering education.

Systems engineer: Nonetheless, many secondary school students have little interest in science or math. Due to scuttlebutt and peer pressure boys especially fond of these subjects are ridiculed; girls are discouraged and made uncomfortable if they take these supposedly masculine courses.

Coach: Allowing girls to join what have traditionally been boy-only teams helps break down barriers to their participation in technical curricula.

School teacher: Administrators and teachers must do more to motivate STEM pursuits to garner more and better trainees capable of addressing world problems. School-wide recognition of accomplishments in these subjects might help. Happily, there are a growing number of STEM programs.

Atheist: European gymnasiums have enjoyed the reputation of being rigorous and almost university level. Why hasn't the US achieved this? Could it be that our children are coddled, being immersed in their parents' comfortable lifestyle?

Parent: Our lifestyle is not that comfortable, and we sacrifice for our kids!

Enterprise systems engineer: But again, more must be done in K-12 to lay the groundwork; by high school it is too late.

Systems thinker: The problem *is* that fundamental.

6 Experience/maturation

Facilitator: We've considered cultural background, and upbringing and education as foundational influences. Professional life begins when one's schooling is over and it's time to get serious. This leads to the last two topics, experience and maturation. So moving right along, let's examine a typical engineering career.

6.1 Career start and realities

Systems engineer: A new engineer may not even consider SE, because they are so embroiled in their specialty. Some branch out to cover, explore, or at least dabble in other areas. Lucky ones receive mentoring from a senior colleague, or maybe collectively from fellow employees. A newbie engineer learns to respect the abilities and contributions their colleagues. Self-doubt is an obstacle to overcome until one's own contributions blossom. An engineer quickly learns the rules of the game and assesses the limits of legitimate and approved actions.

More aggressive actions can be advocated but prudent management decisions should not be challenged.

Systems thinker: Engineers learn that teams have periods of forming, norming, storming, and performing; continuous high performance is unrealistic.

Boss: Good bosses challenge employees with minimum goals but encourage them to go beyond routine expectations. The grace period associated with new assignments should be reasonable but not too generous.

Follower: A new engineer should be trained to understand a problem through first-hand interactions, a great way to appreciate complex nuances [5]. Experience under several bosses may be invaluable. Leadership/management styles should be observed carefully to form one's own persona or modus operandi.

Boss: There are rules of cultural behavior in an organization, both spoken and unspoken. The new engineer is expected to conform, at least until they prove their worth. Nonconformity may be tolerated if they're good enough technically.

System of systems engineer: An employee who does not conform is likely to be marginalized, demoted, or fired. Choice assignments, good raises, and promotions are strong incentives to conform and perform.

Complex systems engineer: Most SE is embedded in projects or programs run as businesses. The main objectives are obtaining funding, ensuring a profit, and pleasing the customer/sponsor. There are lots of politics. Hopefully, the outcome of the effort will serve a purpose that falls within the public interest. However, many participants are motivated just by technological challenge and recognition for doing well.

Enterprise systems engineer: Watching leaders/managers operate in this environment, and ascertaining what motivates them is enlightening. Many are just looking to keep their nose clean, get a raise or bonus, avoid being demoted, and being able to preserve some personal time. Others build a successful track record, receive higher salaries and promotions, and rise to the top of their organization or pursue enviable assignments elsewhere.

6.2 Role models and feelings about work

Enterprise systems engineer: Most engineers view their boss as a potential role model. Assuming the boss is doing the right kinds of things, s/he will be emulated, and supportive underlings will help make them look good. Otherwise, employees may seek more worthy authorities.

Complex systems engineer: Commendably, a good leader does the right things, even though mistakes are sometimes made. Good managers tend to do things right, and this is fine as long as the right things are being pursued! One may latch onto a mentor and take advantage of their guidance, perhaps following them up the chain of command. Alternatively, one may prefer the challenge of not accepting any special favors from management, and insisting on making the grade alone, speaking truth to power, fighting a broken system, and trying to do the right instead of expedient thing.

Liberal supporter: It's good to think about what has been accomplished. Ask yourself: Is what I'm doing really making a difference, or am I actually part of the problem? One may be working on communications, and not building nuclear weapons, but what if the messages control the delivery of those bombs?!

Systems engineer: Engineers are often classified as antisocial or nerdy. This is consistent with some of their behavior, although engineers typically get along well together. Since systems engineers must interact with others in fulfilling their responsibilities, bosses will appreciate any social skills they exhibit.

Boss: You're right. It's really ideal if I can rely on employees to cooperate collectively in doing what I mandate.

Conservative advocate: What is your reaction when problems arise? Do you deal with them yourself as much as possible? If not, do you try practical fixes as learning experiments and ask for help, as appropriate? If that is unsuccessful, do you go to the boss and complain, or do you bring him/her solutions?

Facilitator: Again, thank you for your participation and stimulating thoughts. Our last meeting topic is maturation.

6.3 Spiritual and career maturity

Atheist: If the religion taught to you as a child becomes harder to believe, one might be nudged toward agnosticism. This process might be accelerated after observing lots of people accept horrible events, citing God's will; or, even believing God is all-powerful, blame nature for hurricanes, tsunamis, and earthquakes, etc.

Complex systems engineer: There may be a maturity parallel in thinking about SE. Being taught about SE life cycles with all their accoutrements, and then observing your project/program team following these procedures without success, what does one conclude? Wouldn't one doubt traditional SE and consider other methodologies?

Agnostic: In any event, a process of maturation that establishes spiritual and professional balance in life is usually accomplished by mid-career. Then what?

Lover: Most engineers find out what they're really good at by their 30s or early 40s. These fellas and gals receive suitable recognition for their skills and accomplishments, not only from their bosses but from their colleagues, as well.

Boss: That is not a time to rest on laurels. An engineer should redouble technical efforts to have even more impact. Considerably more latitude in choosing which problems to attack may be afforded to those good enough.

6.4 Building on career accomplishments

Complex systems engineer: Seeking collective understanding, collaboration, and self-organization are principal tenets of effective engineering-team practice.

Enterprise systems engineer: Sharing information iteratively and bilaterally is crucial for gaining mutual trust. New engineers will find this leads to greater opportunities for collaboration. Unfortunately, many organizations over-protect and withhold highly useful information, and sharing is often punished.

Conservative advocate: So? Information sharing should be done only with permission from the boss.

Boss: Otherwise, it can be too damaging to our efforts. Engineers must be sensitive that any blatant missteps are not too detrimental to recipients. As in any family, this can hurt the parent organization.

Systems engineer: These lessons will be important in a project team in becoming dedicated to solving problems without personally attacking colleagues.

Enterprise systems engineer: Engineers should have considerable latitude in how they do their jobs if they're around awhile; some missteps can be excused.

System of systems engineer: Having provided enough technical guidance, engineers need to know when to close the deal, i.e. learn to be good salespeople!

Follower: Late career can be the best time of all. One can review what was learned and accomplished and reapply that wisdom in new contexts.

Lover: Mentoring others is another important activity. One of life's greatest satisfactions is to pass on wisdom acquired through hard work, experience, and a history of honest and trusting relationships.

Enterprise systems engineer: Receiving unexpected, unsolicited compliments from former employees about how they liked working for you is also satisfying.

Complex systems engineer: One never truly fails unless one gives up. Learn from mistakes (made by yourself but preferably others); nothing would get done if nothing was attempted. So take informed risks and pursue opportunities.

System of systems engineer: Resuscitate unfinished work as it better applies.

7 Conclusion

Facilitator: We have some time remaining for brief comments. Are there any?

Theist: Thanks to much of the discussion, I'm becoming even more of a believer in systems engineering!

Systems engineer: Me, too! Maybe there's more to this than I thought, as well, and I'll cogitate on that while re-thinking from where I came.

Systems thinker: Summarizing, Table 2 offers our perspectives on SE attributes to help a systems engineer characterize her/his approach and performance. One could ask oneself which cells of Table 2's 3 × 3 matrix apply to them and to what relative extent. If a particular cell dominates, is that appropriate? What changes might be made to become more effective as a systems engineer? Similarly, Fig. 1 suggests how several categories of a system engineer's existence are proportioned at each stage of their lives. One could contemplate how these trends have affected their past SE behaviors, and how possible personal re-directions might improve future SE accomplishments.

Complex systems engineer: I think these descriptions fit my career and current mindsight very well. However, I will try not to be judgmental about anyone else's interpretation. Objectively, any chosen way is really no better than any other possibility. As we must realize by now, it all depends on a multitude of circumstances in each of our lives. Thinking through what we have done, are doing, and want to do in SE is worthwhile. I recommend such an exercise!

Facilitator: I think these remarks sum things up pretty well. If there are no more comments, I thank you all for participating, and best wishes in solving more of our most difficult problems through improved SE practices.

References

[1] Gorod, A., White, B.E., Ireland, V., Gandhi, S.J., & Sauser, B.J., *Case Studies in System of Systems, Enterprise Systems, and Complex Systems Engineering*, CRC Press, Taylor & Francis Group: Boca Raton, FL, 2015.
[2] Raymo, C., *When God is Gone Everything is Holy – The Making of a Religious Naturalist*, Sorin Books: Notre Dame, IN, 2008.
[3] White, B. E., Managing uncertainty in dating and other complex systems, *Conference on Systems Engineering Research (CSER)*, 15–16 April 2011, Redondo Beach, CA.
[4] Checkland, P., *Systems Thinking, Systems Practice – Soft-Systems Methodology: a 30-year retrospective*, Wiley: New York, NY, 1999.
[5] Giuliani, R.W., *Leadership*, Miramax Books (Hyperion): New York, NY, 2002.

Value-sensitive design for indigenous people of Oaxaca, Mexico

D. E. Santos-Reyes[1] & A. J. Mendoza-Jasso[2]
[1]*I'chi Research and Engineering, Mexico*
[2]*Industrial Engineering Department,*
Mixteca University of Technology (UTM), Mexico

Abstract

This paper describes an ongoing research and development project on engineering value-sensitive design for indigenous people called Mixtecs, and for wider segments of the system of focus, in Oaxaca, Mexico. Although the Mixtec people have creatively maintained their fundamental way of doing things from the pre-Hispanic past to the present day, they are being influenced by a new wave of conflicting factors. These factors, namely economical, technological, environmental and global issues, threaten the survival of the Mixtecs' approach. In addition to these issues, the fast growing rate of crime also impedes growth. A crucial way to understand the current Mixtecan situation is to use complex systems thinking to address the core issues that inhibit growth and subsequently to implement action. It is believed here that there are essential issues of value and these have implications for engineering design. Engineering value-sensitive design may have important implications in addressing the Mixtec people's situation without compromising their values, culture and traditions.
Keywords: innovation, design, sustainability, crime, value.

1 Introduction

The Mixtecs have creatively maintained their fundamental way of doing things from the pre-Hispanic past to the present day. However, they are being influenced by a new wave of conflicting factors over the last few decades. These include environmental, economical, technological and global issues that threaten their survival approach. In addition to these, the fast growth of crime rate is also a major obstacle for motivating growth [1]. Complex systems approach [2, 3] is

www.witpress.com, ISSN 1755-8336 (on-line)
doi:10.2495/978-1-78466-155-7/019

used as the means to develop an understanding of the current situation of the Mixtecs. Cleary, there are essential issues of value and these have implications for engineering design. This demands the consideration of all the conflicting factors that impend the existence of the way of these indigenous people. In other words, engineering value-sensitive design may have important implications in addressing the Mixtecan economic and technological advance without compromising their values, culture and traditions. The relevance of innovation, engineering and value-sensitive design, as well as the disruption of the Mixtecs' way, are discussed in Section 2. Section 3 presents the current situation and the research approach adopted for addressing the problem-solving process. Section 4 highlights the main focus of the authors' efforts to address the Mixtecs' situation and, finally, some relevant conclusions and further work are presented in Section 5.

2 Innovation, engineering and value-sensitive design

2.1 Responsible innovation

Innovation, according to Schumpeter's classical concept [4], involves the creation of new and improved products and processes, the use of existing technology into new applications, the discovery of new resources and the opening of new market segments. It is widely recognised nowadays that this is a world characterised by continuous and fast development of new technologies. Technology is, however, never value neutral [5]. When new technologies emerge and diffuse, there is a presumption that there will be changes in both nature and society. These changes are the result of a dynamic process of continuous conjunction between technological change and social change.

It seems that there are two ways to advance an understanding of how technology innovation is affecting and will affect the future. First, the positive impact of the innovation on economic growth, industrial productivity and organisations' competitiveness has been extensively acknowledged and documented over the last few years. The other way is that there has been a rapidly developing interest in issues such as sustainable design [6, 7] and renewable energy technology [8]. Furthermore, recent debate has focused on the questions of responsible innovation (RI) [9]. The development and adoption of new technologies enable changes in social organisations and values. This is done by creating new possibilities and increasing the number of options available to - people [10]. This requires creativity to address an increasing number of values, such as privacy, security, inclusion, sustainability, democracy, safety, accountability, to mention a few. This is the concern of RI. It is about developing technology and designing products that integrate customers/users' values.

2.2 Disruption of the Mixtecs' way (I'chi)

Recent studies suggest that the Mixtecs can be classified as native Americans that arose from an ancestral Asian population which migrated to the Americas more than 21,000 years ago [11]. The country of these indigenous people lies within the states of Guerrero, Puebla and chiefly in Oaxaca. As shown in Fig. 1, the territory occupied by the Mixtecs extends eastwards from the Pacific coast of Mexico into the high mountain country of the interior. In Oaxaca, they occupy a province known as the Mixtec region, which in turn consists of highland, lowland and coastal sub-regions.

Archaeological, ethnographical and anthropological studies suggest that pre-Columbus Mextecs had the state-of-the-art technology to deal with survival. They grew their own food and relied on rain to irrigate their soil. Agricultural techniques helped to protect, conserve and nourish the cultivatable soil. Small-scale irrigation systems such as wells, dams and canals were used to divert seasonal flood-water to irrigate fields in the piedmont. Food production revolved around the *milpa*, invented and diffused more than 9,000 years ago. This system was, and still is today, a response of not only the Mixtecs, but of all indigenous people of Mexico, to environmental, social, economic and cultural factors. The milpa consisting mainly of corn, squash and bean plants represents a point in – evolutionary and adaptive complex processes. Along with this technological progress, corn tortilla and related processing technology were developed throughout Mexico. A tortilla is a multifunction soft, thin, flat, circular-shaped bread made from finely ground corn flour. This core platform technology further developed and diversified into a great number of new products. Today, this is the main constituent of the Mexican diet. In addition to the milpa, a wide variety of other sources of sustenance included plants, such as the prickly pear fruit from the nopal cactus, avocados, cacao, chilli, green tomato, maguey and ceremonial plants such as native tobacco. Mixtecs also possessed the skills for free-range growing turkeys. In parallel with other cultures of the world, they discovered death and since then they invented tools for killing. In relation to the provision of food, the people of the rain hunted edible wildlife.

Figure 1: Mixteca region, Oaxaca, Mexico.

Indigenous inhabitants of the Mixtec region had very well established designs and production systems of domestic artefacts [12]. They have designed and produced crafts such as pottery, lithic, baskets, amongst other products. Mixtecan technologies were closely linked to the natural world. They wore modest clothing, usually made of 'maguey' fibre, and lived in houses made from thatch and wattle-and-daub, sometimes with stone or adobe foundations. Similarly, herbal medicine had been used within the Mixtecan culture to prevent and treat diseases. This systemic approach to health sourced the healing materials mostly from within their own geographical location. The main sources of energy were solar, biomass and oil. Although the amount of produced energy was small and costly, they knew how to convert one form of energy into another form to support production of sustenance.

When they transformed from nomad to sedentary, Mixtecs developed sophisticated architectural systems, such as pyramids and squares. Usually, these systems incorporated visual art features, such as carved stones, paintings and figurative ceramics. Similarly, they had a sophisticated approach to communicate: a language of their own, including pictorial texts. The latter disrupted by the conquistador, whilst the former is still spoken, but is evaporating slowly. As discussed elsewhere [13], the graphical communication evolved from incidental petroglyphs and rock paintings to a sophisticated system of pictographic manuscripts in the form of screen-fold books, known as codex, of deerskin or paper, and painted cotton cloth.

2.3 Value-sensitive design

Under the influence of a barbaric invasion, issues beyond their control, such as religious, social, economic, technological and health changes drove Mixtecs to adopt creatively some elements of the European culture without disrupting totally their own way. Anyone who has actually observed the Mixtec people will realise that the fundamental way of doing things continues to exist from the pre-Hispanic past to the present day. Recent research on the present day of indigenous people of Oaxaca agrees with this statement. However, the people of the rain, cloud or heat, as they are known in the original language, are being influenced by a new wave of conflicting factors namely environmental, economical, technological, criminal and global issues. This ancient civilisation has not been conquered from without but is being destroyed by itself from within. This demands the consideration of all the conflicting factors that threaten the existence of the way of the Mixtecs. However, the often conflicting, sustaining self-identity and adopting globalisation issues is not the only tension Mixtecs encounter. It is believed here that there are essential issues of value. Furthermore, these concerns have implications for design. In other words, engineering value-sensitive design may have important implications in addressing the conflicting issues of the survival of the Mixtecs. What follows – outlines the approach to these issues.

3 The Mixtecs' current situation

3.1 The approach

Figure 2 unfolds the complexity of the system of focus. It intends to portray the Mixtecs as a whole, a complex adaptive system, instead of as a sum of self-contained people and culture. The approach seeks to understand better how this totality evolves through the millennia. A crucial way to exhibit the current Mixtecan situation is to think from complex systems thinking. This is, to think of Mixtec people as inextricably involved with other people, near and far, in a global community interaction. It attempts to frame the current situation in order to focus thinking on the very core issues that give rise to policy and decision makers, the Mixtecs themselves, and researchers to think, and subsequently to implement action. The purpose is to frame the factors, such as ecological, physical and technological, to understand the observable rhythms and structures of the Mixtecs' survival. Additionally, these factors describe interactions with geography, culture and socio-technical circumstances. The use of this systemic description of the problem situation and any associated remedy of this gives rise to new considerations that have previously been unnoticed or deliberately disregarded in favour of the established status quo. It is customary of public problem solvers, in the guise of delivering better solutions to indigenous people's situations, to overlook the systemic nature of the problem, which translates into problem creation. This very consideration of neglecting the complexity of the indigenous people's situations should now be brought onto the focus point in order to eradicate this reductionist action.

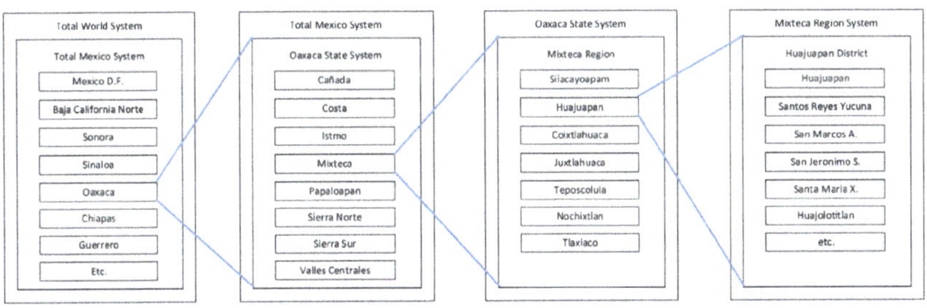

Figure 2: The Mixtecs recursive levels of being.

3.2 The situation

3.2.1 Mixtecs and the World

It seems that a defining feature of the Mixtecs' unease over the last few decades has been globalisation. The interconnection of world economies has driven global trade, consumerism, technology and knowledge flows that are profoundly influencing Mixtecs' way. This is a great tension between the Mixtec people and the world people. The Mixtec people should be able to maintain an independent existence within a global world. As the Mexican, and the world, population inexorably increase, the natural environment continues to deteriorate and the gap between the rich and the poor increases, the people of the rain face the dilemma of whether to keep favouring quality of life or shift to standard of living above all other values. Clearly, it is important for the Mixtec people to value and improve themselves, and it is equally important for the Mixtecs to contribute in the improvement of the Mexican society within the global community.

3.2.2 Cooperation and competition

The current global economic interconnectedness is unparalleled and is becoming more pronounced with each passing year. Globalisation has increased competition not only between companies that contend in the same market place but also between countries and individuals. As the Mexican nation builds into a global economy, successful competition is paramount. On the other hand, it is well known that ancient civilisations, including Mixtecs, are highly skilled at working together towards the same end. In this sense, Mixtecs can contribute to finding balance between competition and cooperation to enable growth.

3.2.3 Old and new

Information and communication technology is homogenising the world people's cultures. Clearly, the world people benefit from this commonality. However, the new is replacing the delightful ancient culture or way of the Mixtecs, for example. There is the need to recognise the Mixtecs' old way that makes it special.

3.2.4 Growth and decay

Forerunners of the modern Mixtecs designed and developed tools to enable them to hunt, gather and later grow food. They crafted stone cutting tools and arrow-heads amongst other inventions. Later, they further developed and diversified these technologies. However, it seems that their ability to innovate was halted by the process of the Spanish invasion. Now, Mixtecs are required to disrupt this decay process by design in order to stimulate growth.

4 Engineering value-sensitive design

4.1 The engineering problem

Survival is the consequence of a number of factors and long periods of evolution. As emphasised above, Mixtecs are involved in highly complex environments, as shown in Fig. 2. They must adapt not only to their harsh local environment but also to the chaotic global environment, whilst retaining their identity. Figure 3 shows a simplified proposed system. The fast analysis system technique, as part of the value analysis, is a rigorous process for designing engineering systems [14]. It is used here for designing complex social systems. The two vertical lines shown in Fig. 3 define the scope of the design problem. The survival of the Mixtecs is the highest level function and it is the consequence of the basic functions, as shown to the immediate right of the left scope line. To sustain an independent existence is a higher level function that represents our proposed approach for the survival of the Mixtecs. The other functions represent the means to achieve this higher level function. These functions were established by the how–why logic. How do Mixtecs sustain an independent existence? It seems that the answer is by producing their own sustenance. Why do Mixtecs produce their own sustenance? Because they need to sustain an independent existence. The functions shown in Fig. 3 were arranged by comparing each other using the how–why logic.

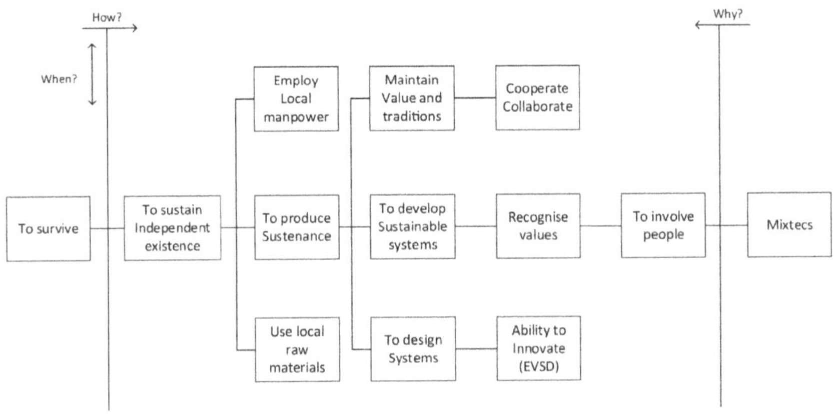

Figure 3: Function analysis system technique for the Mixtecs' survival.

The practice of engineering design is the means by which the authors intend to improve the Mixtecs' current situation in a number of ways, including systems such as agricultural or construction. In particular, value-sensitive design is used to facilitate our ability to innovate engineering systems (Fig. 3). As society strongly believes that medical practice is likely to make one into a healthier person, engineering too has the capability for moral and value formation. However, engineering process is not automatic any more than it is automatic for medical practice. As value and engineering design are closely related, to engage both requires the ability to innovate and to participate in practical reasoning. To achieve an independent existence, Mixtecs must develop and produce their own sustenance and services in the same way they used to do in the pre-Spanish era. The following section highlights some relevant development projects that intend to comply with the function: to develop sustainable systems shown in Fig. 3.

4.2 Development projects

To meet the demand on the Mixtecs' performance, novel solutions and concepts are being investigated and developed by the authors. A coherent analysis and synthesis of systems, such as energy sources and conversion, engineering food products and processes, architectural systems and agricultural systems, is paramount to ensure success. As depicted in Fig. 4, integration of ingenuity, science and technology is the core of our teleological mechanism of novelty. In addition, the optimum performance of systems can only be designed by using advanced engineering practice. Various engineering systems are under development. These include machines to process round-shaped native natural materials, such as reed, sticks and otate. This may increase significantly the productivity of the production of baskets of diverse shape and size, which are traditional Mixtec products. Regarding architectural systems, a system to dampen earthquake vibration has been conceived, as reported elsewhere [15], and a demonstration house will be built over the next 2 years.

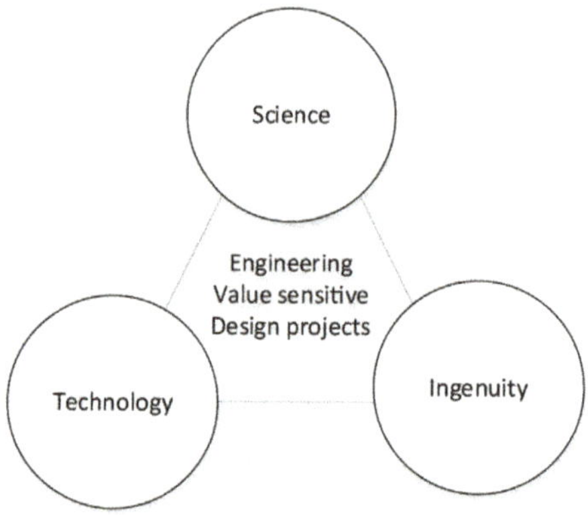

Figure 4: I'chi approach to research and engineering projects.

5 Conclusions

This article argued that the indigenous people of Oaxaca, Mexico, called Mixtecs are experiencing a new wave of factors that affect their growth. The authors, with the hope of improving this situation, use RI, in particular, value-sensitive design. A number of development projects are underway, which will be implemented in the small village of Santos Reyes Yucuna, Oaxaca, Mexico, over the next few years. The successful implementation and outcomes of these projects, by providing the knowledge and skills to this people, may go some way to reducing both the rate of emigration and poverty, including the likelihood of their involvement in crime – be that as criminal or victim.

References

[1] Santos-Reyes, D.E. & Santos-Reyes, J.R., Patterns of temporal diffusion of crime in Mexico. *International Journal of Safety and Security Engineering*, **2(2012)**, pp. 54–68, 2012. doi: http://dx.doi.org/10.2495/safe-v2-n1-54-68

[2] Beer, S., *Brain of the Firm*, John Wiley & Sons: Chichester, 1995.

[3] Rouse, W.B., Health care as a complex adaptive system: implications for design and management. *The Bridge*, **38(1)**, pp. 17–25, 2008.

[4] Schumpeter, J.A., *Business Cycles: A Theoretical, Historical, and Statistical Analysis of the Capitalist Processes*, Macmillan: New York, 1939. doi: http://dx.doi.org/10.1522/030021081

[5] Winner, L., Do artefacts have politics? *Daedalus*, **109(1)**, pp. 121–136, 1980.

[6] Porter, M.E. & van der Linde, C., Green and competitive: ending the stalemate. *Harvard Business Review*, September–October, pp. 120–134, 1995.

[7] Hart, S.L., Beyond greening: strategies for a sustainable world. *Harvard Business Review*, January–February, pp. 66–76, 1997.

[8] Pehnt, M., Dynamic life cycle assessment (LCA) of renewable energy technologies. *Renewable Energy*, **31**, pp. 55–71, 2006. doi: http://dx.doi.org/10.1016/j.renene.2005.03.002

[9] Owen, R., Bessant, J. & Heingtz, M. (eds), *Responsible Innovation: Managing the Responsible Emergence of Science and Innovation in Society, Wiley: Chichester, UK, 2013.*

[10] Manders-Huits, N., What value in design? The challenge of incorporating moral values into design. *Science and Engineering Ethics*, **17**, pp. 271–287, 2011. doi: http://dx.doi.org/10.1007/s11948-010-9198-2

[11] Zerylnick, C., Torroni, A., Sherman, S.L. & Warren, S.T., Normal variation at the myotonic dystrophy locus in global human populations. *American Journal of Human Genetics*, **56**, pp. 123–130, 1995.

[12] Balkansky, A.K. & Croissier, M.M., Multicrafting in prehispanic Oaxaca. *Archeological Papers of the American Anthropological Association*, **19(1)**, pp. 58–74, 2009. doi: http://dx.doi.org/10.1111/j.1551-8248.2009.01013.x

[13] Jensen, M.E.R.G.N. & van Broekhoven, L.N.K. (eds), *Mixtec Writing and Society*, Knaw Press: Amsterdam, 2008.

[14] Miles, L.D., *Techniques of Value Analysis and Engineering*, McGraw-Hill: USA, 1972.

[15] Santos-Reyes, D.E. & Lawlor-Wright, T., Understanding design in nature can benefit product conceptual design. *International Journal of Design & Nature and Ecodynamics*, **4(2)**, pp. 105–122, 2009. doi: http://dx.doi.org/10.2495/dne-v4-n2-105-122

Influences in a biologically complex adaptive system: environmental stress affects dental development in a group of Romano-Britons

A. H. Brook[1,2], K. S. B. Koh[1] & V. K. L. Toh[1]
[1]*School of Dentistry, University of Adelaide, Australia*
[2]*Institute of Dentistry, Queen Mary University of London, England*

Abstract

The human dentition shows the general characteristics of a complex adaptive system. Interactions at a molecular level of genetic, epigenetic and environmental factors lead to the emergence of tooth germs. Subsequently differentiated specialised cells control mineralisation and then mature teeth emerge, showing marked variation. The mature dentition provides a record of developmental influences from its initiation at 6 weeks in utero to its completion at 20 years of age. It is, therefore, a valuable paradigm of general development.

The substantial collection of skeletal material of 200–400 AD Romano-Britons from Poundbury, Dorset, has been extensively studied, providing much evidence concerning lifestyle and medical conditions. In this study, we investigate further findings concerning dental development. The aims are to compare the findings with those for Modern Britons and to identify the effects of major factors known to affect the Romano-Britons in order to consider the possible synergism of these effects. The findings were that the patterning of the dentition, its sexual dimorphism and the types of developmental defects were similar to modern Britons. However, the Romano-Britons had more anomalies, generally smaller tooth crowns and roots, and more severe enamel defects. Thus, the Romano-Britons' dentition showed evidence of insults occurring at all stages of tooth development. These effects, occurring over a long period of time and affecting tissues with different genetic backgrounds, probably arise from ongoing major environmental insults known to be present in Poundbury. These insults have been identified as excess lead ingestion, poor nutrition and recurrent

WIT Transactions on State of the Art in Science and Engineering, Vol 90, © 2016 WIT Press
www.witpress.com, ISSN 1755-8336 (on-line)
doi:10.2495/978-1-78466-155-7/020

infections. We propose that these factors, acting synergistically, constituted marked environmental stress affecting development.
Keywords: complex adaptive system, dental development, Romano-British dentition.

1 Introduction

1.1 The dentition as a complex adaptive system and model for general development

Investigating human development as a complex adaptive system has the potential both to enhance the understanding of the aetiology of development anomalies and to identify targets for their prevention. We have provided evidence that general and craniofacial developments are complex adaptive systems using dental development as a paradigm [1]. The detention is a valuable model as it is accessible and can be studied non-invasively. In contrast to many other physical and mental aspects of development, the teeth are durable hard tissues that can be accurately and repeatedly measured. The mature dentition provides a robust record of influences during development from its initiation at 6 weeks in utero to its completion at 20 years of age. There are extensive data concerning the genetic/epigenetic/environmental factors involved in dental development from animal and human studies. The genes, including homeobox genes, expressed in the early morphogenetic stages of dental development are common to the development of a number of other tissues. In the later stages of differentiation and mineralisation, different genes control the process and these processes act as a model for general mineralisation.

Dental development (Fig. 1) commences with bottom-up self-organisation as, at a molecular level, the genetic, epigenetic and environmental factors interact. These interactions occur within cells and their surrounding matrices, leading to the development and organisation of a specific number of higher level organs, the soft tissue tooth germs [1]. These emergent structures bear no resemblance to the precursor molecules. Within the tooth germs, cellular multiplication and apoptosis determine their size and shape at specific locations within the dental arches. Subsequent differentiation within the tooth germ provides specialised cells that control the subsequent mineralisation of the tooth germs. The mineralised mature teeth erupt and form a dynamic system that continues to adapt during function.

During development, cells multitask carrying out multiple functions: the enamel forming cells control the deposition of matrix and then its gradual removal as mineralisation proceeds. Self-adaptation occurs as epigenetic switches are used to turn genes on and off and control their expression; environmental factors are involved both directly and by causing epigenetic changes. There are critical phases or tipping points with thresholds determining further development. The performance of this complex adaptive system is improved by diversity with variation in outcome increasing the robustness of the system and allowing evolutionary development in response to environmental

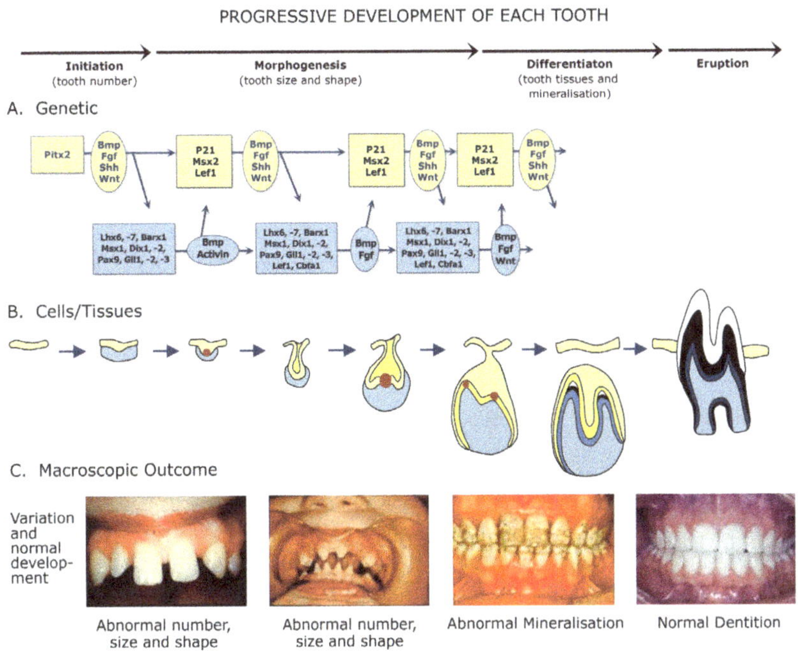

Figure 1: Overview of dental development: upper layer indicates ectoderm in (A) genes and (B) cells/tissues. The shaded lower layer in (A) and (B) indicates mesenchyme. Sections (A) and (B) are modified from http://bite-it.helsinki.fi/ (accessed 10 April 15). Figure reproduced with permission from Australian Dental Journal [1].

changes. Dental development also shows a series of network models; random network, threshold, scale-free, hierarchical and gene regulation network models, with subgraphs, motifs and motif clusters [1]. The erupted mature dentition continues to function as a complex adaptive system, with changes due to tooth wear and local environmental insults leading to adaptive responses that retain function.

1.2 The Romano-British collection

The substantial collection of skeletal material in the British Museum (Natural History) London from the excavation of the AD 200–400 cemetery in the Poundbury camp of the civitas capital of Durnovaria, Dorchester, Dorset has been extensively studied, providing much evidence concerning their lifestyle and the medical conditions they suffered [2]. The population was of native British origin and is thought to have lived as a close knit society with limited immigration for five or six generations. Age and sex determinations of the skeletal material were made by the staff of the British Museum using long bones, pelvic girdles and skulls. The people of Poundbury Camp probably had an agricultural lifestyle with cereal as the predominant element in the diet [2]. From an examination of 364 non-adults from Poundbury Camp, Lewis [3] provides evidence of high levels of rickets, scurvy and trauma in the children: iron deficiency and vitamin C and D deficiency are likely to have occurred together in the malnourished children. The presence of typical bony lesions is suggestive also of megaloblastic anaemia due to vitamin B_{12} deficiency. In addition, gastrointestinal infections occurred [4]. The level of malnutrition in Poundbury Camp is marked and in some cases may be related to the maternal health during pregnancy [3]. The teeth and bones of the Poundbury population have been found to have a high lead content [5–8], with one potential source of this lead being the drinking vessels [2].

2 Aims

The aims of this paper are to evaluate findings concerning dental development from this Romano-British population in comparison to results for Modern Britons in order to investigate the influence of the major environmental factors present in Poundbury over the duration of the full developmental process and to consider possible synergism of these effects.

3 Materials and methods

This investigation examined data for four outcomes of dental development: tooth number; crown size and shape; root size and shape and enamel structure. A total of 486 skulls had intact jaws and dentitions and were examined. Within this sample, subsamples were used for different aspects of the study according to the suitability of the material for scoring. For crown size, the subsample number was determined by statistical calculation. In each table, the sample size is given.

Visual examinations were made for tooth number and for enamel hypoplastic defects. Manual measurements using dial calipers were undertaken for mesio-distal and bucco-lingual crown size. Customised wooden wedges were constructed to position the skulls reproducibly and enable standardisation of six radiographs of each skull to cover the full dentition. Using these radiographs, the number of teeth formed in each individual, including unerupted teeth, was confirmed and the root number, size and shapes were scored. The criteria for diagnosis followed Brook [9] and Brook and Smith [10] and the hand measurement technique followed by Morrees *et al.* [11]. The same criteria were applied by the investigators in the studies of modern Britons used for comparison [9, 10, 12]. For each parameter, the levels of reproducibility of scoring were high.

4 Results

The results are summarised in Tables 1–4. Table 1 indicates the findings for tooth number and visual size and shape; Table 2 for crown mesio-distal size measurements; Table 3 for root size and shape and Table 4 for enamel hypoplastic defects.

Table 1: Prevalence for anomalies of tooth number, size and shape.

	Romano-British (n = 486), %	Modern British (n = 1,115), %	Ref.
Hypodontia of third molars	39	12.7	[13]
Hypodontia excl. third molars	13	4.4	[12]
Microdontia excl. third molars	6.4	2.5	[12]
Supernumerary teeth	1.2	2.1	[12]

Table 2: The mean mesio-distal diameters of each permanent tooth type and their standard deviations in 14- to 24-year-old Romano-Britons.

Tooth	Male (n = 30)		Female (n = 30)		M:F differences	
	Mean	SD	Mean	SD	*t*-test	*p*-value
Maxilla						
Central incisor	8.42	0.30	7.87	0.34	6.6437	<0.0001
Lateral incisor	6.57	0.40	6.43	0.37	1.4073	0.1647
Canine	7.60	0.32	7.28	0.24	4.3818	<0.0001
First premolar	6.48	0.35	6.26	0.29	2.6511	0.0103
Second premolar	6.16	0.43	6.06	0.28	1.0674	0.2902
First molar	9.90	0.45	9.45	0.48	3.7461	0.0004
Second molar	9.40	0.59	9.10	0.60	1.9527	0.0557
Third molar	8.78	0.95	8.58	0.79	0.8866	0.3790
Mandible						
Central incisor	5.11	0.25	4.96	0.25	2.3238	0.0237
Lateral incisor	5.77	0.32	5.53	0.37	2.6872	0.0094
Canine	6.68	0.31	6.30	0.21	5.5587	<0.0001
First premolar	6.65	0.34	6.37	0.30	3.3823	0.0013
Second premolar	6.62	0.35	6.50	0.38	1.2722	0.2084
First molar	10.99	0.49	10.50	0.34	4.5000	<0.0001
Second molar	10.48	0.37	10.16	0.55	2.6441	0.0105
Third molar	10.58	0.77	10.20	0.14	2.6594	0.0101

Table 3: Prevalences in skulls and sex distributions of root anomalies.

	No. of affected males	Prevalence	No. of affected females	Prevalence	Sex ratio (M:F)
Three-rooted mandibular first molar	3	2.3% ($n = 132$)	3	2.4% ($n = 127$)	1:1 (NS)
Fused root (upper and lower molar teeth)	13	8.5% ($n = 153$)	29	19.5% ($n = 149$)	1:2.2 ($p < 0.02$)
Cuneiform root (upper and lower molar teeth)	19	12.4% ($n = 153$)	38	25.5% ($n = 149$)	1:2 ($p < 0.01$)
Taurodonism (lower molars)	48	34.8% ($n = 138$)	31	22.3% ($n = 139$)	1:0.67 ($p < 0.05$)

Table 4: Prevalence of enamel hypoplastic defects [10].

	Romano-British (n = 178) (%)	Modern British (n = 1,518) (%)
Individual prevalence	37.1	14.6
Tooth prevalence	8.0	1.7
Percentage with ≥4 teeth affected	25.0	1.3
Frequency of horizontal hypoplastic grooves	29.8	1.8

These results show that in comparison with Modern Britons of European ancestry, the Romano-Britons exhibited;

1. The same types of dental development defects affecting tooth number, crown size and shape, root size and shape and enamel structure.
2. The same sexual dimorphism in tooth size and dental developmental defects with females having smaller teeth and a higher prevalence of hypodontia while males have larger teeth and a higher prevalence of supernumerary teeth.
3. A threefold greater prevalence of hypodontia (congenitally missing teeth) and microdontia (abnormally small teeth with reduced shape).
4. Smaller crown size.
5. Smaller root size with reduced shape. There were highly significant associations between fused and cuneiform roots ($p < 0.001$) and both were associated with hypodontia ($p < 0.002$ and $p < 0.05$).
6. A threefold greater prevalence of enamel hypoplasia with the majority of affected individuals having four or more teeth involved.

5 Discussion

Dental development has been shown to have the general characteristics of a complex adaptive system [1]. The findings of this investigation are that the dentition of these Romano-Britons showed the same general characteristics as that of Modern Britons and can therefore be considered as a complex adaptive system, developing from the interactions of genetic, epigenetic and environmental factors. However, there were substantial differences in the dimensions of teeth and the prevalence of anomalies between the findings for Romano-Britons and the Modern Britons.

This is modelled in Fig. 2 for tooth number, size and shape. This diagram is a development of the unifying aetiological model proposed by Brook [12] from the findings on Modern Britons and other Europeans. It is based on the normal distribution of tooth size, with separate curves for males and females reflecting the sexual dimorphism in tooth size. Thresholds are superimposed on these curves defining anomalies of tooth number, size and shape. The anomalies have the characteristics of quasi-continuous variables of multifactorial determination [12]. Supporting this model, a number of studies have shown that females have a higher prevalence of hypodontia and microdontia, whereas males have a higher prevalence of megadontia and supernumerary teeth [14]. In addition, there are statistically significant associations between hypodontia and smaller teeth and between supernumerary teeth and larger teeth [15–17]. Plotting the curves from the Romano-British data shows them to the left of the Modern British curve, reflecting the findings of higher frequency of hypodontia and microdontia,

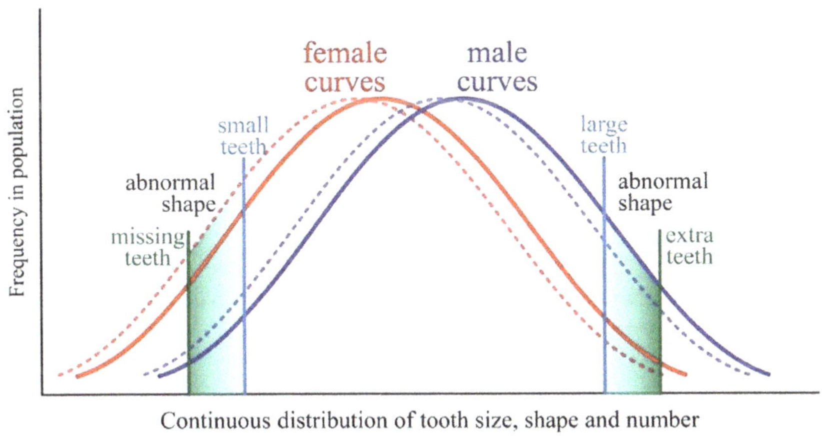

Figure 2: The unifying aetiological model of Brook [12] is developed here to include changes in tooth shape as well as tooth size. The solid lines are for Modern Britons and the dotted lines are for Romano-Britons. Figure reproduced with permission from Australian Dental Journal [14].

smaller crown and root size and lower frequency of supernumerary teeth for this Poundbury population.

The next consideration is what aetiological factors influenced this complex adaptive system to produce these differences (Fig. 3). During morphogenesis, the number, size and shape of the teeth are determined by a series of interactions between the genes in the epithelial and mesenchymal tissues (Fig. 1) [1]. The soft tissue tooth germs emerge from these reiterative interactions of the same genes at molecular and cellular levels. This explains the epidemiological and clinical findings of associations between anomalies of tooth number, size and shape, as in the Romano-Britons and Modern Britons that is modelled in Fig. 2.

However, in the subsequent differentiation stage within the soft tissue tooth germs that leads to mineralisation and the mature tooth, different genes are switched on and control the processes of matrix deposition and calcification [18]. In the Romano-Britons, the mature teeth show much greater levels of enamel defects than the Modern Britons (Table 4) and so they have a substantially higher level of development defects throughout the whole developmental period. It is unlikely that this is due to genetic mutations as these would need to be multiple, affecting the different genes involved in morphogenesis and in differentiation. Probably, the factor or factors causing the differences are generalised environmental agents present in the population throughout the whole period of dental development.

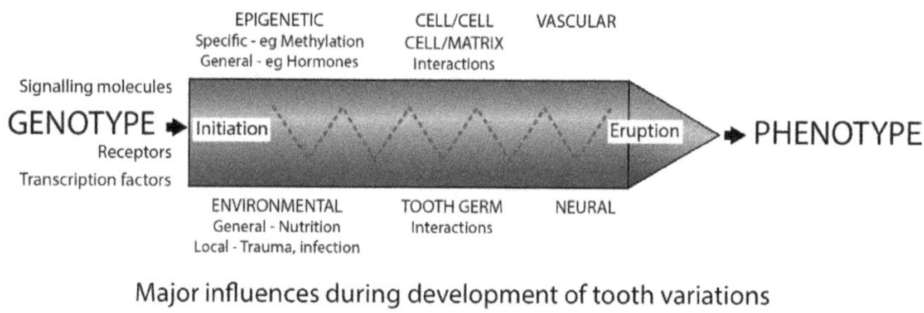

Figure 3: The major influences on dental development from genotype to phenotype. In the Romano-Britons, three major environmental factors have been identified which may have part of their effect through epigenetic changes. Figure reproduced with permission from Australian Dental Journal [14].

In Section 1 of this paper, evidence has been briefly reviewed for three areas of major, long-term environmental insults known to affect development in the Poundbury population: high levels of lead ingestion; poor general nutrition with identified vitamin and mineral deficiencies; and recurrent infections. Each of these environmental insults has been suggested as a cause of dental developmental defects in animals and human populations [19, 20] and has been identified as present in Poundbury [2]. Rather than having individual effects, they may have acted in summative or even synergistic ways to constitute major environmental stress throughout dental development in the Romano-British population. While some of the action of these factors could be direct effects on the molecular and cellular processes of tooth development, they may also act indirectly through influences on the immune and endocrine systems and the production of epigenetic changes. If this is so, there could be transgenerational epigenetic changes affecting subsequent generations [21].

These findings have implications for modern populations suffering major environmental stress and emphasise the value of considering dental development both as a complex adaptive system and as a valuable model for evaluating factors affecting general development.

Acknowledgements

This investigation was supported by the Australian Dental Research Fund. We are grateful to Theya Molleson of the Natural History Museum, London, who provided not only access to material but also valuable advice and support. Contributions to the database used were made by a series of postgraduate students: L.K. Foo, C.A. Johns, M. Scheers, J.M. Smith and C.A. Underhill.

References

[1] Brook, A.H., Brook O'Donnell, M., Hone, A., Hart, E., Hughes T.E., Smith, R.N. & Townsend, G.C., General and craniofacial development are complex adaptive processes influenced by diversity. *Australian Dental Journal*, **59(1 Suppl)**, pp. 13–22, 2014.

[2] Farwell, D.E. & Molleston, T., *Poundbury – Vol 2 – The cemeteries*, Friary Press: Dorchester, UK, 1993.

[3] Lewis, M.E., Life and death in a civitas capital: metabolic disease and trauma in the children from late Roman Dorchester, Dorset. *American Journal of Physical Anthropology*, **142(3)**, pp. 405–416, 2010.

[4] Walker, P.L., Bathurst, R.R., Richman, R., Gjerdrum, T. & Andrushko, V.A., The causes of porotic hyperstosis and criba orbitalia: a reappraisal of the iron-deficiency-anaemia hypothesis. *American Journal of Physical Anthropology*, **139(2)**, pp. 109–125, 2009.

[5] Waldron, H.A., Mackie, A. & Townshend, A., The lead content of some Romano-British bones. *Archaeometry,* **18(2)**, pp. 221–227, 1976.

[6] Waldron, H.A., Khera, A., Walker, G., Wibberley, G. & Green, C.J., Lead concentration in bone and soil. *Journal of the Archaeological Society*, **6**, pp. 295–298, 1979.

[7] Stack, M.V. & Whittaker, D.K., Lead, zinc and calcium concentrations in Romano-British teeth. *Journal of Dental Research*, **61(4)**, pp. 563, abstract 244, 1982.

[8] Whittaker, D.K & Stack, M.V., The lead, cadmium and zinc content of some Romano-British teeth. *Archaeometry*, **26(1)**, pp. 37–42, 1984.

[9] Brook, A.H., Variables and criteria in prevalence studies of dental anomalies of number, form and size. *Community Dentistry and Oral Epidemiology*, **3(6)**, pp. 288–293, 1975.

[10] Brook, A.H. & Smith, J.M., The aetiology of developmental defects of enamel: a prevalence and family study in East London, U.K. *Connective Tissue Research*, **39(1–3)**, pp. 151–156, 1998.

[11] Morrees, C.F.A., Thomsen, S.O., Jenson, E. & Yen, P.K-J., Mesiodistal crown diameters of deciduous and permanent teeth in individuals. *Journal of Dental Research*, **36(1)**, pp. 39–47, 1957.

[12] Brook, A.H., A unifying aetiological explanation for anomalies of human tooth number and size. *Archives of Oral Biology*, **29(5)**, pp. 373–378, 1984.

[13] Shinn, D.L., Congenitally missing third molars in a British population. *Journal of Dentistry,* **4(1)**, pp. 42–44, 1975.

[14] Brook, A.H., Jernvall, J., Smith, R.N., Hughes. T.E. & Townsend, G.C., The dentition: the outcomes of morphogenesis leading to variations of tooth number, size and shape. *Australian Dental Journal,* **59(1 Suppl)**, pp. 131–142, 2014.

[15] Brook, A.H., Griffin, R.C., Smith, R.N., Townsend, G.C., Kaur, G., Davies, G.R. & Fearne, J., Tooth size patterns in patients with hypodontia and supernumerary teeth. *Archives of Oral Biology,* **54(Suppl 1)**, pp. S63–S70, 2009.

[16] McKeown, H.F., Robinson, D.L., Elcock, C., Al-Sharood, M. & Brook, A.H., Tooth dimensions in hypodontia patients, their unaffected relatives and control group, measured by a new image analysis system. *European Journal of Orthodontics,* **24(2)**, pp. 131–141, 2002.

[17] Khalaf, K., Robinson, D.L., Elcock, C., Smith, R.N. & Brook, A.H., Tooth size in patients with supernumerary teeth and a control group measured by an image analysis system. *Archives of Oral Biology,* **50(2)**, pp. 243–248, 2005.

[18] Brook, A.H., Multilevel complex interactions between genetic, epigenetic and environmental factors in the aetiology of anomalies of dental development. *Archives of Oral Biology,* **54(Suppl 1)**, pp. S3–S17, 2009.

[19] Small, B.W. & Murray, M.J., Enamel opacities: prevalence, classifications and aetiological considerations. *Journal of Dentistry*, **6(1)**, pp. 33–42, 1978.

[20] Chen, H.S., Tsai, Y.C., Chen, K.K. & Hsu K.J., Detrimental effects of maternal lead exposure during pregnancy and lactation on molar development in the young rat. *Bulletin of Environmental Contamination and Toxicology*, **89(2)**, pp. 240–244, 2012.

[21] Zambrano, E., Marinez-Samayoa, P.M., Bautista, C.J., Deás, M., Guillén, L., Rodrigues-González, G.L., Guzmán, C., Larrea, F. & Nathanielsz P.W., Sex differences in transgenerational alterations of growth and metabolism in progeny (F_2) of female offspring (F_1) of rats fed a low protein diet during pregnancy and lactation. *Journal of Physiology*, **566(1)**, pp. 225–236, 2005.

Section 4
Multi-agent applications

A multi-agent solution for managing complexity in English to Sinhala machine translation

B. Hettige, A. S. Karunananda & G. Rzevski
*Department of Computational Mathematics,
University of Moratuwa, Sri Lanka*

Abstract

Machine translation turns out to be an inherently complex process requiring serious attention to morphological, syntactic and semantic complexity within both the source and the target languages. Most of the existing approaches to machine translation (MT) circumvent the complexity with the assumption that morphological, syntactic and semantic analysis can be done independently and sequentially. This has resulted in depriving us of the opportunity to use the language complexity to generate high-quality translations. In view of this, research has been conducted to develop a multi-agent systems solution for MT that uses the language complexity as an opportunity for generating a more realistic translation from English to Sinhala. This multi-agent solution primarily comprises a six-agent swarm to deliberate on morphological, syntactic and semantic concerns of the source and the target languages without being constrained to operate in a sequential manner. These agents use the ontology of corpora and dictionary of two languages. This approach is inspired by the fact that people understand a sentence by incrementally reading through words while simultaneously considering the syntax and semantics. As such, when the system progresses in identification of words one by one, both syntactical and semantic concerns are entertained up to the current point of reading. As a result, initially decided words may be changed due to the present concern of morphology, syntax and semantics. A translation system has been implemented on the multi-agent system development framework named MaSMT. Experiments show that the multi-agent solution for MT gives promising results for translating sentences

WIT Transactions on State of the Art in Science and Engineering, Vol 90, © 2016 WIT Press
www.witpress.com, ISSN 1755-8336 (on-line)
doi:10.2495/978-1-78466-155-7/021

of an average length and further research has been carried out to accommodate translation of long sentences.
Keywords: machine translation, multi-agent systems, Sinhala language.

1 Introduction

Machine translation (MT) has been a research concerning the translation of text or speech from one natural language (source language) to another natural (target language) [1]. According to the complexity of the natural languages, MT has become a reality in research since last few decades. At present, a number of approaches are available to MT, including human-assisted, rule-based, statistical, example-based, knowledge-based and hybrid. Using these approaches thousands of MT systems have been developed to translate natural languages around the world. Among these approaches, statistical approach has been the widely used successful approach for the MT since last decade. Google translator [2], Moses [3], SYSTRAN [4] are the successful statistical-based MT systems in the world. In addition to the above, rule-based approach is powerful and commonly used approach for MT especially for target languages that have different writing methods for the written and spoken. Further, rule-based systems are capable to provide grammatically correct translation by using morphological, syntactical and semantic rules of the source and target language. Number of MT systems have been designed through rule-based approach, including Apertium [5], GramTrans [6], Anusaaraka [7] and BEES [8].

In view of the inherent complexity of the natural languages, MT itself turn out to be a complex task requiring analysis and generation of morphological, syntactic and semantic information of both source and target languages, respectively. Furthermore, a successful MT system requires providing suitable solutions to handle translation issues such as word ambiguity, idiomatic phrase translation, different syntax structures, translate anaphors and co-reference. Much of existing approaches to MT circumvent the complexity with the assumption that morphological, syntactic and semantic analysis can be done independently and sequentially. This has resulted in depriving of opportunity to use the language complexity to generate high-quality translations. As a result, at present, even the most successful MT systems in the world could not achieve > 90% accuracy.

In responding to above issue, a research has been conducted to develop multi-agent systems solution for MT that uses the language complexity as an opportunity for generating a more realistic translation from English to Sinhala. This multi-agent system solution primarily comprises six-agent swarms to deliberate on morphological, syntactic and semantic concerns of the source and the target languages without being constrained to operate in a sequential manner. These agents use the ontology of corpora and dictionary of two languages. The multi-agent approach to MT is inspired by the fact that people understand a sentence by incrementally reading through words while simultaneously considering the syntax and semantics. Stated another way, a person never reads all the words in a sentence first, and looks for the grammatical accuracy and the meanings of the sentence at last. Thus, our solution operates in a unique manner where a morphological agent deliberates with syntactic and semantic agents to identify the words incrementally in a sentence. As such, when the system progresses in identification of words one by one, both syntactical and semantic concerns are entertained up to current point of reading. Accordingly, initially decided words may be changed due to the present concern of morphology, syntax and semantics.

Concurrently with the reading through the words in an English sentence, corresponding Sinhala words are also identified from the bilingual database. The initially identified Sinhala words may change when Sinhala syntactic and semantic agents start executing to form a grammatically correct sentence in Sinhala. The translation happens only after reading an English sentence fully.

The system has been implemented with multi-agent development framework MaSMT, which is specially designed to develop MT applications through the multi-agent approach. Translation system consists of six sub-systems, namely English morphological system, English syntactical system, English semantic system, Sinhala morphological system, Sinhala syntactical system and Sinhala semantic system that handle morphology, syntax and semantics of both languages. The English morphological agents are responsible to analyze the English morphology and Sinhala morphological agents are responsible to generate Sinhala word for the given morphological information. English syntax agents are responsible to analyze the syntax structure of the English sentence and Sinhala syntax agents are responsible to generate Sinhala syntax structure for the given syntax information. Semantic agents are also capable of analyzing and identifying the semantic information as required for the translation. Final solution provides through the communication among these sub-systems of agents.

The rest of the paper is organized as follows. Section 2 gives a brief summary of the existing approaches, systems and issues for the MT. Section 3 presents our novel approach to MT. Section 4 explains how system works in practice. Section 5 concludes the paper with a note on further work.

2 Machine translation approaches, systems and issues

This section briefly describes existing approaches, systems and common issues in MT.

2.1 Approaches to machine translation

A number of approaches are available for MT including direct translation, dictionary-based, human-assisted, statistical, example-based, rule-based and hybrid [1]. By using these approaches, a large number of MT systems are developed for many natural languages.

Direct translation is a primitive approach to MT that replaces the words in source language with words in the target language by using a bilingual dictionary. This approach does not care to do linguistic analysis or processing and uses only a bilingual dictionary for the translation. According to this approach, dictionary lookups may be done with or without morphological analysis or lemmatization. In contrast, dictionary-based systems are commonly used for cross-language retrieval systems and ideally suitable for the translation of long lists of phrases on related language pairs, which has same syntactic and structure. Telugu to English translation [9], Japanese–English dictionary-based MT system [10], Bengali and Hindi to English cross-language text retrieval [11] systems are dictionary-based MT systems.

Human-assisted MT approach is particularly used for Indian families of MT. The human-assisted approach uses human interaction for the pre-editing, post-editing and/or intermediate editing stages [10]. This approach uses human support for the semantic handling in the MT. This type of MT system is also called semi-automated MT systems. Statistical MT approach is by far the most widely studied MT method in the field of natural language processing. This approach tries to generate translations using statistical methods based on bilingual text corpora. Moses [12], Google Translator [2], are the popular Statistical MT Systems.

Example-based MT (EBMT) approach uses bilingual corpus with parallel texts as the knowledge base. These types of MT systems are more suitable for the languages with less resources. Example-based systems are trained through the bilingual parallel corpora, which contain sentence pairs. Kyoto-U [13] and example-based English to Sinhala machine translator [14] have used this approach to MT.

The rule-based approach is yet another approach for MT. This approach gives grammatically correct translation by using a set of rules. Rule-based system consists of morphological, syntax and semantics rules to process the translation. Most of these MT approaches run with in sequence and start with morphological analysis of the source language and end with the target language morphological generation. Anusaaraka [7], BEES [8] are some example for the rule-based translation.

The hybrid MT systems use combine method in two or more approaches especially rule-based and statistical MT approaches. English to Malayalam MT [15] and TransEasy [16] system use this approach to MT.

2.2 Existing English to Sinhala machine translation systems

Google translation is one of the popular free multilingual translation services provided by Google. At present, it supports more than 90 languages, including Sinhala. This translation service offers a web interface, mobile interfaces for Android and iOS, and an API for developers who can use to build browser extensions, applications and other software. [17] Sinhala translation is available on Google since December 2014. According to Google translation, its service limits the number of paragraphs and the range of technical terms that can be translated. However, Google translator helps the reader to understand the general content of a foreign language text and it does not always deliver accurate translations.

BEES is another rule-based MT system, powered by theory of varanageema (conjugation) in Sinhala language [8]. BEES system consists of seven modules namely English morphological analyzer [18], English parser [19], English to Sinhala translator, Sinhala syntax generator, Sinhala morphological generator and Transliteration module [20]. English to Sinhala MT system has been implemented with the use of SWI-Prolog, Java and Prolog Server Pages (PSP).

Liyanapathirana and Weerasinghe have developed a Statistical MT system for English to Sinhala [21]. Silva and Weerasinghe have also developed an example-based English to Sinhala MT [14]. This research has attempted to translate government documents in English to Sinhala through a database of examples.

2.3 Common issues in machine translation

Process of the MT can be primarily realized through morphological, syntactical and semantic level concerns of natural languages. As such, in the outset, an MT system should be able to handle morphological, syntactical and semantic issues correctly to provide successful translation. The first and commonest problem in morphological level has been recognized as the ambiguity in words in a sentence [22]. The meaning of the word is imprecise or open to more than one interpretation or some words have multiple interpretations are the main reasons for the word ambiguity. To avoid the word ambiguity, sufficient context or explanation is required [23]. Translating an idiomatic phrase is another common issue on the MT. An idiom is a common word or phrase with a culturally understood meaning that differs from what its composite words' denotations would suggest [24]. To translate the idiomatic phrases, translation system requires to identify these sets of words as a single unit. Therefore, it is difficult to handle large idiomatic phrases.

A sentence having different syntactical structures is a major issue in syntax level concerns of MT. To handle this syntax-level ambiguity, syntax analysis is required. In addition to that, some structures of the target language differ from the source language (e.g. English has SOV order and Sinhala comes with SOV structure. At that point translation is not easy.

Further, a grammatical term for pronoun, which refers back to another word or phrase, known as Anaphora [22] is another issue at the syntax level. This in fact results in semantic-level analysis too. To translate anaphora, MT systems are required to identify the antecedent of the anaphora by matching the features of the anaphor with the nouns of the previous in the nominative form. Then it is required to identify anaphora of the target language. In addition to the above, it is required to translate anaphor from source language to target language; the features of anaphora of source language mapped to the anaphora of the target language. If more than one entry is available for the anaphora in the bilingual lexicon, then the GNP features are matched to get correct anaphora [25].

Co-reference is one of the problems of semantic translation, and it occurs when two or more expressions in a text refer to the same person or thing. To handle co-reference in the translation, complex index is required.

Discourse is another issue of the semantic translation. To handle these semantic issues, it is needed to analyze the discourse structure. In addition to the above facts, some source sentences do not have single correct answers. In the MT point of view, it is an issue to handle through the translation.

3 Design: multi-agent approach to machine translation

The approach behind the MT is based on the hypothesis that words employ as the building block of natural language understanding and MT process requiring attention to morphological, syntactic and semantic on both source and target languages. This is valid for people who read a sentence word by word or otherwise by locating selected words such as nouns and verbs. Further, consequently, meaning of a sentence is determined by the interaction among words, which draw from all aspects of morphology, syntax and semantics, as appropriate. This solution has been designed as a multi-agent system and implemented through the multi-agent development framework MaSMT [26]. Figure 1 shows the architecture of the MaSMT framework.

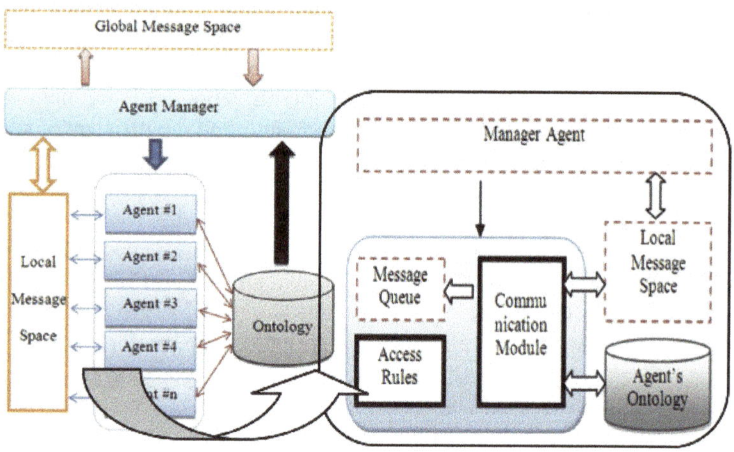

Figure 1: Architecture of the MaSMT framework.

The MaSMT framework is especially designed to develop MT application through the multi-agent platform. The MaSMT framework provides two types of agents, namely ordinary agents and manager agents. The manager agent consists of a number of ordinary agents within its control. Further, manager agents can directly communicate with other manager agents (through the message transport agent) and each and every ordinary agent in the swarm is assigned to a particular manager agent. An ordinary agent in a swarm can directly communicate only with the agents in its own swarm and its manager agent. The MaSMT framework has been implemented by using JAVA and multi-agent system capabilities that have been tested with several applications [27].

English to Sinhala MT system consists of six sub-systems (agent swarms), namely English morphological system, English syntactical system, English semantic system, Sinhala morphological system, Sinhala syntactical system and

Sinhala semantic system. In addition to the above, message transport agents and translator agent are the supporting agents. Figure 2 shows the design diagram of English to Sinhala MT system. According to the multi-agent architecture, each subsystem works as an agent swarm. A brief description of each multi-agent swarm is given below.

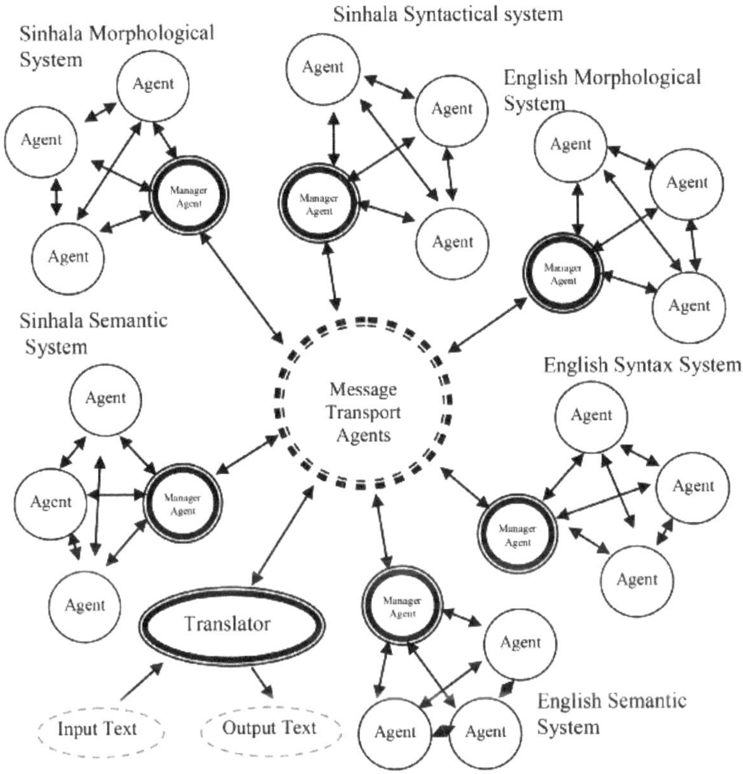

Figure 2: Design of the multi-agent-based English to Sinhala machine translation system.

The English morphological sub-system works as an English morphological analyzer and it is capable to analyze English words and provides the morphological information for the given English word [28]. The morphological system is also a multi-agent system that can communicate with each other agents and provides morphological information as required. The morphological system consists of a manager agent (MaSMT manager) and numbers of morphological agents (agent in the MaSMT) to analyze English morphology. The present system consists of 19 morphological agents to represent 19 English morphological rules.

The English syntax system is the English syntax analyzer of the MT system, which is capable to syntactically analyze the given text. The system also consists of a manager agent and a number of syntactical agents to represent syntax rules.

These agents can identify syntactical category of the input words such as noun phrase, verb phrase, subject, verb, etc. After analysis, the system generates the syntax structure of a given sentence of phrase. English semantic system is also capable to analyze the semantics of the given English text. Semantic system gets support from English–Sinhala bilingual dictionary and Internet to collect semantic information as required.

Sinhala morphological system is worked as Sinhala morphological generator and it generates appropriate Sinhala word for the given morphological information. This system also consists of 132 Sinhala morphological agents to support morphological generation through the Sinhala morphological rules [8]. These grammar rules are based on Sinhala word conjugation forms [8]. The Sinhala semantic system also provides semantic information for the Sinhala words that are generated by the system. This system also uses bilingual dictionary and the Internet to get the required information.

Message transport agent is a supporting agent of the translation system that helps to transport messages for each manager. According to the MaSMT architecture each agent in the swarm can communicate only for each agent in its own swarm and its manager agent. Managers can communicate only with its agents and other managers. A message passing through the manager agents is done through the message transport agents. The translator is a manager agent that reads input sentence form the graphical user interface (GUI) and provides output for the GUI.

According to the multi-agent architecture, each sub-system is capable to communicate with each other and makes their job as required [29]. The system especially does not give any order to start its task first; as well as some systems may wait until other systems finish their task (syntactical system waits until the morphological system completes its tasks). English to Sinhala MT is done through communication of each of the six sub-systems.

4 How the system works

This section briefly describes how MT system works for the given input sentence. According to the translation approach discussed in the paper, translation system works as how a person (non-English person with expertise in target language) translates an English sentence into Sinhala. Most of the professional translators translate a sentence through the two phrases. At first they read all words in the source sentence word by word and get an idea about context and semantics. Then they generate grammatically correct target language sentence with considering the source and target language grammar (especially syntax). This translator uses this mechanism to translate English sentence into Sinhala.

As a first step of the translation, translator agent reads an English sentence from GUI to translate. Then translator reads the sentence word by word on left to right. Assume translator reads a sentence 'good boy reads books.' Then translator gets the first word 'good' and starts the translation process with the support of each sub-system. At this point, all six sub-systems get the word 'good' and start their job. English morphological system analyzes the English word 'good' and sends the morphological information. The English semantic system identifies corresponding Sinhala words for the given English word 'good.' However, English syntax system can start its task after English morphological system finishes its job. Note that, Sinhala morphological system and Sinhala syntax system wait until translator completes the reading of the English sentence.

After that, the translator reads the next word 'boy.' At this point, the English syntax system identifies 'good boy' as a noun phrase and the English semantics system searches for suitable Sinhala words for 'boy' considering the previously selected Sinhala word for good.

Incrementally system reads the word 'reads' as the next word and English syntax system identifies the word 'reads' as a verb phrase. (Note that at this level, English syntax system does the analysis only a phrase level) In this point system consists of a noun phrase 'good boy' and the verb phrase 'reads.' Now English semantic system identifies suitable Sinhala words for the verb 'reads' with considering the previously selected Sinhala words for 'good boy.' As such, system reads all words until the end of the given sentence. According to this approach, this process is same as the person who reads the English sentence and gets some idea about the context. Up to this point, English syntax system does only the phrase-level analysis and semantic system identifies suitable Sinhala words for the each phrase.

When the system progresses in identification of words one by one, both syntactical and semantic concerns are entertained up to current point of reading and initially decided words may be changed due to the present concern of morphology, syntax and semantics.

After reading all the words in the English sentence English morphological sentence identifies the syntactical category of the input English sentence such as subject, verb, object, predicate, etc. Then with communication of English and Sinhala syntactical systems, suitable Sinhala syntax is generated for the given English sentence. Sinhala sentence has a different structure than the English sentence. Therefore, syntactical transform is required to generate the grammatically correct Sinhala sentence. After that, Sinhala morphological system generates the suitable Sinhala word for each Sinhala words considering all the required grammar. Finally GUI system reads the generated target sentence.

5 Conclusions and further work

This paper has reported our research on the use of multi-agent system technology to implement English to Sinhala MT. This solution postulated that MT can be done going through words in a sentence by incrementally handling morphological, syntactical and semantics concerns of both source and target languages. This goes beyond the sequential approach to processing of sentences to MT in conventional systems. Our system has been implemented with six sub-systems namely the English morphological system, English syntactical system,

English semantic system, Sinhala morphological system, Sinhala syntactical system and Sinhala semantic system. English to Sinhala multi-agent-based MT system is a complex system that handles more than 500 agents, which are interconnected through the six swarms. These sub-systems are not centrally controlled and work as required for the translation. Agents in the system have a degree of autonomy but their behaviour is always subject to certain language-specific rules. This MT system has been implemented on the multi-agent system development framework named as MaSMT. It is evident from the experimental results that our translation system generates promising results when translating sentences with average length and further research are being conducted to accommodate translation of long sentences.

References

[1] Hettige, B. & Karunananda A.S., Existing Systems and Approaches for Machine Translation: A Review, *Proc. of the 8th Annual sessions on Sri Lanka Association for Artificial Intelligence (SLAAI)*, pp. 34–40, 2011.

[2] Google Translate Blog, googletranslate.blogspot.com

[3] Moses, www.statmt.org/moses/

[4] SYSTRAN, www.systransoft.com

[5] Forcada M.L., Tyers F.M. & Ramírez-Sánchez G., The Apertium machine translation platform: five years on, *Proc. of 1st Int. Workshop on Free/Open-Source Rule-Based Machine Translation*, pp. 3–10, 2009.

[6] Wiechetek L., Rule-based MT approaches such as Apertium and GramTrans, Online. uit.no/Content/84556/mt.pdf

[7] Chaudhury S., Rao A. & Sharma D.M., Anusaaraka: An expert system based machine translation system, *Proc. of 2010 Int. Conf. on Natural Language Processing and Knowledge Engineering (NLP-KE),* pp. 1–6, 2010.

[8] Hettige B. & Karunananda A.S., A Computational grammar of Sinhala for English-Sinhala machine translation, *Proc. of 2011 Int. Conf. on Advances in ICT for Emerging Regions (ICTer),* pp. 26–31, 2011.

[9] Prasad T.V. & Muthukumaran G.M., Telugu to English Translation using Direct Machine Translation Approach., *International Journal of Science and Engineering Investigations.*, **2(12)**, pp. 25–32, 2013.

[10] Okumura A. & Hovy E., Building Japanese-English dictionary based on ontology for machine translation, *Proc. of workshop on Human Language Technology*, pp. 141–146, 1994.

[11] Mandal D., Dandapat S., Gupta M., Banerjee P., & Sarkar S., Bengali and Hindi to English Cross-language Text Retrieval under Limited Resources, *in Working Notes for the CLEF 2007 Workshop*, 2007.

[12] Koehn P. & Hoang H., Birch A., Chris C., Federico M., Nicola B., Brooke C., Wade S., Christine M., Richard Z., Chris D., Ondrej B., Alexandra C., & Evan, Moses: Open Source Toolkit for Statistical Machine Translation, *presented at the Annual Meeting of the Association for Computational Linguistics (ACL),* Prague, Czech Republic, 2007.

[13] Nakazawa T., Yu K., Kawahara D., & Kurohashi S., Example-based machine translation based on deeper NLP, *in IWSLT,* pp. 64–70, 2006.
[14] Silva A.M. & Weerasinghe R., Example Based Machine Translation for English-Sinhala Translations, *Proc. of 9th Int. IT Conference (IITC 2008), Colombo, Sri Lanka*, pp. 27–28, 2008.
[15] Nithya B. & Joseph S., A Hybrid Approach to English to Malayalam Machine Translation., *International Journal of Computer Applications* **8(1)**, pp. 11–15, 2013.
[16] Liu Q. & Yu S., TransEasy: a Chinese-English machine translation system based on hybrid approach, *in Machine Translation and the Information Soup, Springer*, pp. 514–517, 1998.
[17] Google Translate, Wikipedia, the free encyclopaedia.
[18] Hettige B. & Karunananda A.S., A Morphological Analyser to Enable English to Sinhala Machine Translation, *Proc. of Int. Conf. on Information and Automation, ICIA 2006*, pp. 21–26, 2006.
[19] Hettige B. & Karunananda A.S., A Parser for Sinhala Language - First Step Towards English to Sinhala Machine Translation, *Proc. of 1st Int. Conf. on Industrial and Information Systems*, pp. 583–587, 2006.
[20] Hettige & Karunananda A.S., Transliteration system for English to Sinhala machine translation, *Proc. of Int. Conf. on Industrial and Information Systems, ICIIS 2007*, pp. 209–214, 2007.
[21] Liyanapathirana J. & Weerasinghe R., English to Sinhala Machine Translation: Towards Better information access for Sri Lankans, *in Conference on Human Language Technology for Development*, Alexandria, Egypt, pp. 182–186, 2011.
[22] Wikipedia: Ambiguous words, Wikipedia, the free encyclopaedia.
[23] Center T., Ambiguity Reduction for Machine Translation: Human-Computer Collaboration, Online. http://homes.cs.washington.edu
[24] Gaule M. & Josan G.S., Machine Translation of Idioms from English to Hindi, *International Journal of Computational Engineering Research (ijceronline.com)*, **2(6)**, pp. 50–54, 2012.
[25] Suryakanthi T., Prasad S., Prasad T.V., Translation of Pronominal Anaphora from English to Telugu Language, *International Journal of Advanced Computer Science and Applications, (IJACSA)*, **4(4)**, pp. 75–79, 2013.
[26] Hettige B., Karunananda A.S. & Rzevski G., MaSMT: A Multi-agent System Development Framework for English-Sinhala Machine Translation, *Int. J. Comput. Linguist. Nat. Lang. Process. IJCLNLP*, **2(7)**, pp. 411–416, 2013.
[27] Hettige B., Karunananda A.S. & Rzevski G., Sinhala Ontology Generator for English to Sinhala Machine Translation, *Proc. of KDU International Research Conference*, Colombo, 2014.
[28] Hettige B., Karunananda A.S. & Rzevski G., Multi-agent System Technology for Morphological Analysis, *Proc. of 9th Annual session on Sri Lanka Association for Artificial Intelligence (SLAAI),* pp. 1–7, 2012.
[29] Rzevski G. & Skobelev P. Managing Complexity, WIT Press, 2014.

Using a multi-agent system for supply chain management

L. C. M. Perera & A. S. Karunananda
Department of Computational Mathematics, University of Moratuwa, Sri Lanka

Abstract

Supply chain management (SCM) is a well-known example of a complex system. Classical computing technologies have shown little success in modeling complex systems. However, a large body of research in multi-agent system (MAS) technology has demonstrated how complex systems can be modeled to generate smart solutions, which could not be done otherwise. We have researched on the design and development of MAS for SCM. In this solution, each phase in the supply chain has been developed as an agent enabling communication, coordination and negotiation among the agents to achieve intended business goals. The study investigated decentralized collaborative planning architecture and agents are attached to different containers of the system. The containers have been implemented using a Java Agent Development Framework (JADE) and consist of diverse methods to support collaboration in the supply chain environment. Agents have different behaviors and their decisions are based on defined ontology. The identified key roles in the supply chain are raw material suppliers, manufacturers, distributors and retailers. They perform autonomous tasks with collaboration to accomplish final customer satisfaction.
Keywords: multi-agent system, supply chain management.

1 Introduction

Complexity in communication in supply chain management (SCM) has been cited by many researchers [2]. In general, the entities operating in a supply chain environment have been provided with limited opportunities for information sharing and collaborative planning. Undisputedly, identification of different

WIT Transactions on State of the Art in Science and Engineering, Vol 90, © 2016 WIT Press
www.witpress.com, ISSN 1755-8336 (on-line)
doi:10.2495/978-1-78466-155-7/022

entities in a business process and working together to achieve intended goals for an organization is an essential requirement. SCM consists of material venders, suppliers, manufacturers, distributors, retailers and customers to work as network of agents to achieve a common goal. There are some developed legacy systems and client server architecture applications that have provided marginal benefit for the effective communication in SCM. Although there are several automated solutions for SCM, which use classical software technologies but the issues remain unsolved. However, a large volume of literature has shown the potential of multi-agent system (MAS) technology to effectively use the power of communication as a problem-solving strategy in complex systems.

Multi-agent technology has been used in many areas but industry applications have taken the earliest advancement of agent technology when compared to others. The features of MAS technology can be simply used to handle the complexity in communication, which plays the vital role in SCM. OASIS is an air traffic control system used in Sydney airport, Australia can be introduced as an example for efficient usage of MAS in communication [1]. It consists of agents and they are created automatically when aircrafts arrive at the Sydney airport. Their behavior is goal directed and reactive to capture real-time constraint into accounts.

Collaborative planning architecture is an inherent solution for effective communication [2]. It improves the accuracy in the decision-making process and also established a link between internal and external entities to increase the performance in SCM. This paper consists of multi-agent-based solution to eliminate communication complexity in supply chain environment.

The rest of the paper is organized as follows. Section 2 presents aim and objectives. Section 3 discusses the current trends and issues of SCM while Section 4 describes multi-agent technology. Section 5 explains our approach to SCM using MAS technology. Section 6 deals with design of SCM followed by discussion on our proposed methodology in section 7. Results of the effective communication are dealt with in section 8. Section 9 describes conclusion and future work.

2 Aim and objectives

Our aim is to critically study the SCM domain with a view to identify current practices and issues. Analyze the existing solution in SCM with the view to define the research problem and possible technology. Complete depth study about multi-agent technology and its applications to design and implement MAS solution for SCM. Finally, evaluate the MAS using a real-world scenario.

3 Current trends and issues in SCM

There are legacy systems that have been developed for SCM [2]. Such systems include advanced planning systems (APS), enterprise resources planning (ERP) and e-commerce systems. Supply chain can be viewed as connected ultimate suppliers and ultimate customers. The suppliers distribute merchandise at the right quantities to the right locations at the right time to minimize the system-wise cost, while satisfying the service-level requirements. In SCM, performance depends on willingness and coordination with others. As a result of this one entity in supply chain is necessarily affecting the performance of the other entity. However, suppliers are inclined to make own decisions in SCM. They involve in many organizations but, sometimes have to depend on decisions made by manufacturers [3].

Effective communication and information sharing improve working capital by 25% [7]. Regardless of the business type in the SCM process, the primary objective of the every business entity is to earn maximum profits by increasing working capital, streamlining accounts, managing debtors and eliminating unnecessary cost. Therefore, the importance of accurate communication mechanism is an essential factor for the SCM. The introduction of agent technology and Java Agent Development Framework (JADE) platform has fulfilled the needs of effective communication mechanism. SCM is regarded as a strategic essential factor in obtaining great deal of competitiveness in the business environment [8]. In SCM if there is one single company exists in one directional way collaboration may not be useful and not complicated. However, cumulative business growth and connection with other companies in the business process agent technology can be introduced as promising candidate in SCM. SCM, network has facilitated different functions of distribution options for procumbent materials, convert it to intermediate and finished goods and finally distribute over-the-end customers [9].

According to the swarm intelligence, collections of agents work like ants or honey bees [10]. Their searching starts from parallel points and each member carries out intended operation to maximize utility function for sharing information with others. As a new technology paradigm for conceptualizing, designing and implementing computer applications and to eliminate potential issues related to current software applications was achieved using agent technology [11]. SCM is a solution to the some industrial issues [1]. Therefore, the bullwhip effect [12] can be identified as a major problem in the domain of SCM. In SCM, a single modification has done by the end customer and it affects to the entire process is named as bullwhip effect. From the industry viewpoint, production planning, inventory management and transport are the main areas of finding significant problems in SCM. However, handling these entities as an individual and to manage them to behave using their own decisions is rather complicated and difficult. For instance, transport and the inventory management are known as N-P hard problems and are very difficult to handle because all of these entities are interdependent. In order to eliminate above constraints needs a system, which supports collaborative planning and information sharing. Object modeling and simulation are new mechanisms that have been developed for SCM [13]. Object modeling have facilitated modeling and individual agents. However, throughout their implementation have not indicated the use of ontologies, message parsing techniques and collaboration for effective communication.

Procedural reasoning system architecture is one of the best architecture of multi-agent technology [14]. It has been deployed in many industry applications not only in SCM but also in other complex commutation systems. However, as a simplest architecture for an agent can be introduced as belief, desire and intention architecture also called as BDI architecture [15] which consists of simplest characteristics, associated for intelligence agents. Supply chain is denoted as complex dynamic system and every member within the system engage the local interactions [16]. MAS have permitted satisfactory results for problems with higher complexity by using classical mathematical model [17]. Development of software agents for distributed data processing can be accomplished by using a framework [18]. They also discussed that growth of contact centers, e-commerce's and value chain reside additional problems of entropies data management and exploitation.

Layered MAS architecture is based on agent commitments [19]. It has illustrated that agents are as components and their interconnections specified as a term of commitments. Combine architecture is an apparent solution for eliminate communication issues in SCM [20]. Today's most of the supply chain functions are static and relay on long-term relationships among trading partners but, more dynamic and flexible practices between entities in SCM provide better solution for customers and suppliers [21]. Introduction of theoretical framework model for collaborative inventory management has effectively benefited successfulness of the SCM [22]. However, the intended results were limited to conceptual and theoretical implementation. To optimize supply chain performance need effective coordination among the entities but due to dynamics of the business operation in an organizational environments are facing difficulties in all aspects. Therefore, sometimes it leads to delayed shipments, production failures, employees are informed as sick and customers change their requirements [23]. In real-world business environments, there are no obligations for any organizations to remain with the supply chain. According to their own judgment, companies can join or leave at any time from the SCM process [24]. In particular, our solutions have implemented using JADE environment and included supply chain functions in real time with more customer involvements for the purpose of wastage reduction, increase overall profits margins and finally to satisfy the customer requirements.

In view of the above, we identify the research problem as the lack of proper coordination and communications in supply chain environment have resulted in malfunctioning of whole process and leading to customer dissatisfaction, profit loss and increase wastage.

4 Multi-agent technology

The concept of MASs has become imperative in artificial intelligence and computer science due to the power of its features and wide industry acceptability. The theory behind agent is, question about what is an agent and the use of mathematical formalisms for representing and reasoning the properties of an agents [25]. However, MAS technology is combination of group of agents sharing information and acting collaboratively to achieve common goals. There are many software projects that have been used in MAS technology to eliminate communication complexity in various domains, including SCM in distributed environment. An agent technology has shown great potential for solving complex distributed problems in over last decades with the growth of inherent distribution. Therefore, decomposition of large-scale systems to individual agents has achieved their common goals by interaction with others [26].

4.1 Features of agents

In a broader sense, agents could be a kind of software or hardware working on behalf of someone with some specific features [27]. Among others, agents are autonomous so that they can operate without the direct involvement of humans and others and retaining some kind of control of their actions and internal state. Agents can also interact with other agent (sometimes humans) via some kind of agent communication language (ACL). The ACL [28] has been proposed by the Foundation for Intelligent Physical Agents as a proposed standard language for agent communications. Knowledge Query and Manipulation Language is another proposed standard [6, 29]. Agents can perceive the environment changes and can respond in a timely manner what has occurred on it. It may be from physical world, change in the graphical user interface, collection of other agents, from Internet and sometimes combination of all. They simply not act in response to the environment but, able to exhibit the goal-directed behavior by taking the initiatives.

5 Our approach to SCM using MAS technology

This section presents our approach by describing the hypothesis, inputs, outputs, processes, features and users of the novel solution. The proposed solution will be able to model complex system for SCM to address the issues of communication among the parties involved in the process [29].

MAS for SCM have been designed to accept multiple inputs coming from different entities of the supply chain process. There are two modes of output generated by the system. These outputs are coming as organizational-related and the customer-related aspects. Having entered the inputs to the system uses MAS technology and ontology to generate the outputs. For the purpose of understanding, we have illustrated the inputs and outputs generated from raw material supplier agents and manufacturer agents. Raw materials, demand from the manufacturer agents, capacity bookings (man, machine and space) are the inputs from the raw material agents while supplied quantities, customer details, delivery schedules, material cost and future forecasting are generated as outputs. The customer demand is primary input of the manufacture agent. However, capacity bookings, delivery schedules and raw material prices and supplier details are the other inputs for the manufacturer agents. Whereas invoices, dispatch details, quality reports and delivery schedules are the generated outputs from the manufacture agents.

The knowledge required for agents to operate is stored in a common domain ontology and personal ontology. Agents basically access their personal ontologies to generate decisions. Our solution has separate agents for selecting suppliers and generates decisions about pricing, discounts, quality of the products, after service and the past records of entities in SCM. However, agents access common domain ontology that has broad description of the entire supply chain process and made available to any agent with define permissions. Inventory management rules are introduced such as FIFO (First in First Out), LIFO (Last in First Out) according to the product category and availability of stocks.

As a promising technology JADE [5] has been used to automate entire SCM process and Eclipse Luna has used as development environment. Overall features of the proposed system are not derived from the inputs and outputs and are basically non-functional requirements of the system. Agents are created using framework therefore each agent is attached to the containers consumes limited resources and produce high performance. Despite of other multi-agent technologies JADE is available as free software component hence development cost is marginal. Ability to perform under limited resource environment, installation and access through the mobile devises increased the rapid growth of multi-agent technology. Instead of the standalone environment the agent is accessible through the web interface with minimum bandwidth.

6 Design of MAS-based SCM

Previous section has described the approach to the SCM, including hypothesis, inputs, outputs, process and features of the system. In this section, the main objective is to present a high-level understanding of the design architecture, modules and relationships among the different entities involved in SCM.

6.1 Two components architecture of SCM

In SCM, information sharing and collaborative planning are the most principal features have been used to eliminate complexity in distributed communication. Therefore, it can be introduced as parallel components and are highly coupled each other in communication.

6.2 Entities involved in SCM

There are four main categories of entities (modules) that have been identified namely raw material suppliers, manufacturers, distributors and retail suppliers. All of these entities access common ontology and their personal ontologies while information sharing and collaborative planning. They are sometimes found as hierarchical within the supply chain process but in most cases they have a parallel behavior.

Figure 1 illustrates who are the responsible entities involved in the SCM process and their dependencies. According to Fig. 1, there are limited number of manufacturer agents that operate in the process and they hold significant responsibility when compared to raw material suppliers. In other words, they have more control over the entire supply chain process. However, retail supplier agents are dealing with the customers and any bullwhip effect from the customers end will lead to substantial change in whole SCM process. Therefore, handling the behaviors of each entity is a demanding tasks, and essential to implement comprehensive knowledge base including domain and personal for smooth functioning of the SCM process.

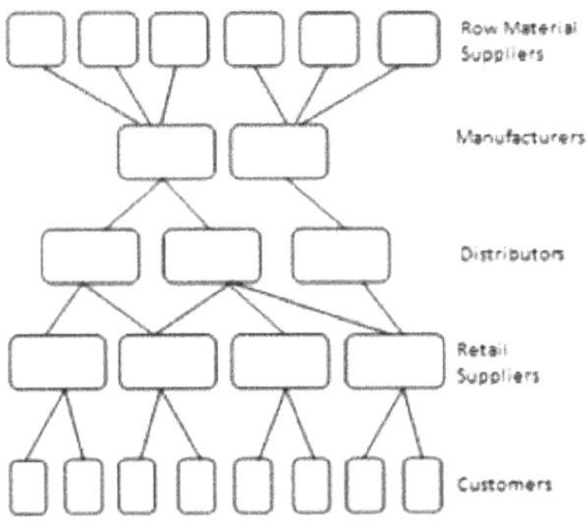

Figure 1: Module-based SCM architecture.

6.3 Relationship among the entities in SCM

Figure 2 describes that agent behaviors, roles and responsibilities. Each agent is attached to different containers and is authenticated by main container of the

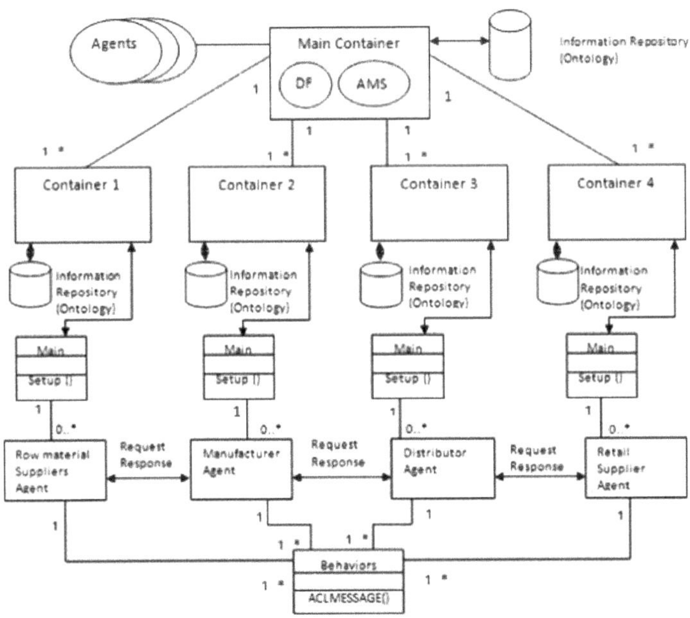

Figure 2: High-level architecture of SCM.

system. The framework we have used to implement containers is described in detail in implementation section [30, 31].

It also illustrates that agents are connected to sub-containers to access their personal ontologies while agents are attached to main container also access domain ontology. Agent behaviors are highly coupled with connected ontologies but, main container has authority for agent administration such as transfer agent from one location to another (agent migration), suspend agent whenever it's necessary and terminate an agent. Agent-to-agent communication occurs via message passing and it is included in the diagram as request and response.

7 Our proposed methodology

To illustrate the proposed solution, we have included evidence of the resulted outputs. There are two modes of ontologies that have implemented in the system namely personal ontology and domain ontology. However, the knowledge base is physically coded in XML files and MySQL database. In order to generate locally collaborative decisions agents have to access MySQL database and for the global decisions they have to access defined knowledge in XML files. In addition to that the defined ACL message passing mechanism has benefited the successful negotiation in SCM [19]. For implementation purpose Eclipse Luna 4.4.0 has used with JADE [5]. The agents are implemented as separate java classes and their initialization also coded in separate java class.

8 Results of the effective communications

Our aim in this section is to shift the emphasis from design and implementation to practice. It is evident that the results obtained from the communication had with retailer agents and distributer agents are coordinated to deliver and receive products in right time with right quantities.

Figure 3 illustrates decisions have taken by raw material agent and manufacturer agent according to the communication and negotiation procedure. Letter 'D' represents the raw material agents while letter 'S' denotes the manufacture agents. The communication process starts sending offers from manufacturer agent to raw material agent to negotiate the price. If both agents have agreed the prices, then they start to negotiate about the supply quantities. When manufacture agents have received the right quantities form according to demand received by retailers the negotiation process will reach to end. Fig. 3 outputs only two categories of agents from SCM domain.

Figure 4 explains that outputs from the end of negotiation process for receiving and offering prices for the certain product category.

According to Fig. 4, 'D1' to 'D5' represents five raw material agents. From the beginning of the communication process D1 received 35 for the quoted 25 and D2 also received 45 for the quoted 35. However, from D3 to D5 have received lower prices when compared to their quoted prices. This has occurred due to competition among raw materials agents because more raw material agents have joined with the negotiation process. Due to this communication

```
 - S1 Offer Best Price D3.  Will answer Rs:47 in 1021 ms
 - S2 Offer Best Price D3.  Will answer Rs:52 in 675 ms
 - S3 Offer Best Price D5.  Will answer Rs:52 in 1394 ms
Proposed quantity is: 32 from S1
Proposed quantity is: 488 from S2
Proposed quantity is: 1349 from S3
Timeout ! quote Rs:44 from S2 is no longer valid

Best Price Rs:13 from S3
ORDER at Rs:42 Got Proposal Rs:42 from D2 & My Price is Rs:13
== AGREE
Current offer is : 42
New price from ontology file: 10
Requested quantity is: 92 from S3 is answer by D2 with proposed quantity of: 1140
Got AGREE from Manufacture Agent:- S3
----------------------------------------- Negotiation Completed -----------------
```

Figure 3: Communication results of raw material and manufacture agents.

Figure 4: Raw material agents requested and received price schedule.

process benefits have received by manufacturer agents because they are able to receive raw materials for competitive prices.

9 Conclusion and future work

The literature review and design section have introduced two areas in supply chain mainly information sharing and collaborative planning. We have stated in literature section by outlining some industry problems and in what way the concept of multi-agent technology addresses the complexity for solving problems.

After that we have moved into the multi-agent technology as novel approach for the SCM. Our solution described that the features and behaviors of agents and power of multi-agent concept for solving problems, which cannot solve by using conventional SCM applications. We also presented that multi-agent architecture and motivation to use MAS technology by comparing other conventional applications. The system can be applied to forecast trends in supply chain functions and to maximize business objective of an organizations and finally satisfy the end-customer requirements. In fact, the results of the effective communication sections clearly indicate that further negotiation has benefited for the significant change in the SCM which cannot achieve by other conventional technologies. However, we believe that the presented model and multi-agent technology shows potential for the future researchers to eliminate communication complexity in distributed environment.

Acknowledgement

L. C. M. Perera would like to thank Mrs. Harindra R. Fernando for her great support and encouragement.

References

[1] Moyaux, T., Chaib-draa, B. & D'Amours, S., supply chain management and multiagent systems: an overview, *Springer Abstract*, **28**, pp. 1–27, 2006.

[2] Hernández, J.E., Poler, R., Mula, J., & de La Fuente, D., A multiagent based-model for the collaborative planning process in decentralized supply chain networks, *XIII Congreso de Ingeniería de Organización*, pp. 1012–1020, 2009.

[3] Swaminathan, J.M., Smith, S.F. & Sadeh, N.M., Modeling supply chain dynamics: a multiagent approach, *Decision Sciences*, **29(3)**, pp. 607–632, 1998.

[4] Kumar, V., & Srinivasan, S., A review of supply chain management using multi-agent system, *International Journal of Computer Science Issues*, **7(5)**, pp. 198–205, 2010.

[5] Um, W., A study of multi-agent based supply chain modeling and management, *iBusiness*, **2(4)**, pp. 333–341, 2010.

[6] Min, J.U., Bjornsson, H. Agent based supply chain management automation. *Proceedings of the Eighth International Conference on Computing in Civil and Building Engineering (ICCCBE-VIII)*, pp. 1001–1006, 2000.

[7] Ojha, M., Optimizing supply chain management using gravitational search algorithm and multi agent system. *Proceedings of the International Conference on Soft Computing for Problem Solving (SocProS 2011) December 20–22, 2011*, pp. 481–491, 2012.

[8] Lau, S. K & Li, X. A multi-agent approach towards collaborative supply chain management. *International Conference on Electronic Business*, B. Liu (Eds.), pp. 929–935. The Chinese University of Hong Kong: Hong Kong, 2005.

[9] Fransoo, J.C., & Wouters, M.J., Measuring the bullwhip effect in the supply chain, *Supply Chain Management: An International Journal,* **5(2)**, pp. 78–89, 2000.

[10] Julka, N., Srinivasan, R., & Karimi, I., Agent-based supply chain management – 1: framework, *Computers & Chemical Engineering,* **26(12)**, pp. 1755–1769, 2002.

[11] d'Inverno, M., Kinny, D., Luck, M., & Wooldridge, M., A formal specification of dMARS. *Intelligent Agents IV Agent Theories, Architectures, and Languages*, Springer, pp. 155–176, 1998.

[12] Pereira, D., Oliveira, E., Moreira, N., and Sarmento, L., Towards an architecture for emotional BDI agents, in *EPIA*, vol. 5, pp. 40–47, 2005.

[13] Zhu, X., Agent based modeling for supply chain management: examining the impact of information sharing, Kent State University, 2008.

[14] Carvalho, R. & Custódio, L., A multiagent systems approach for managing supply-chain problems: new tools and results, *Inteligencia Artificial,* **9(25)**, pp. 79–88, 2005.

[15] Popirlan, C.I., Stefănescu, A. & Stefănescu, L., Multi-agent approach for data analysis in a knowledge-based system for contact centers, *World Academy of Science Engineering and Technology*, **59**, pp. 1126–1131, 2011.

[16] Chopra, A.K., & Singh, M.P., Elements of a business-level architecture for multiagent systems. *Programming Multi-Agent Systems*, Springer, pp. 15–30, 2010.

[17] Srinivasan, S., Singh, S.K., and Kumar, V., Multi-agent system based service oriented architecture for supply chain management system, *International Journal of Computer Applications*, **27(5)**, pp. 12–16, 2011.

[18] Haitham, A., 2010. Applying Electronic Supply Chain Management Using Multi-Agent System A Managerial Perspective.pdf

[19] Fu, Y., Piplani, R., De Souza, R., & Wu, J., Multi-agent enabled modeling and simulation towards collaborative inventory management in supply chains. *Simulation Conference, 2000. Proceedings. Winter*, vol. 2, pp. 1763–1771, 2000.

[20] Rady, H.A., Multi-Agent System for Negotiation in a Collaborative Supply Chain Management, *International Journal of Video & Image Processing* and *Network Security*, **11(5)**, pp. 25–35, 2011.

[21] Chen, Y., Peng, Y., Finin, T., Labrou, Y., Cost, S., Chu, B., Sun, R., & Willhelm, R., A negotiation-based multi-agent system for supply chain management, *Work. Notes Agents*, vol. 99, 1999.

[22] Shehory, O.M., *Architectural properties of multi-agent systems*, Carnegie Mellon University, The Robotics Institute, 1998.

[23] Chen, B., Cheng, H.H., & Palen, J., Integrating mobile agent technology with multi-agent systems for distributed traffic detection and management systems, *Transportation Research Part C: Emerging Technologies*, **17(1)**, pp. 1–10, February 2009.

[24] Wooldridge, M., & Jennings, N.R., Intelligent Agents: Theory and Practice-1995 wooldridge95intelligent.pdf

[25] Labrou, Y., Finin, T., & Peng, Y., The current landscape of agent communication languages *IEEE Intell. Syst.*, vol. 14, no. 2, pp. 45–52, 1999.

[26] Chopra, A.K., & Singh, M.P., Agent Communication, MIT Press, 2011, www.csc.ncsu.edu/faculty/mpsingh/papers/mas/Agent-Communication-chapter.pdf

[27] Foundation for Intelligent Physical Agents, FIPA 97 Specification, Part 2, Agent Communication Language 'fipa_OC00018.pdf

[28] Kumar, S., & Kumar, U., Java agent development framework, *International Journal Research*, **1(9)**, pp. 1022–1025, 2014.

[29] Anandampilai, B., Content-based multicasting using JADE, *International Journal of Soft Computing,* **2(3)**, pp. 422–425, 2007.

[30] Chen, B., Cheng, H.H., and Palen, J., Integrating mobile agent technology with multi-agent systems for distributed traffic detection and management systems, *Transportation Research Part C: Emerging Technologies*, **17(1)**, pp. 1–10, February 2009.

[31] Jennings, N.R., Wooldridge, M., & Sycara, K., A Roadmap of Agent Research and Development, 1998 roadmap.pdf

Design and development of an agent-based model for business operations faced with flood disruption

C. Li & G. Coates
Durham University, UK

Abstract

Small and medium enterprises (SMEs) constitute a major component of the United Kingdom's economy, accounting for 99.9% of all private sector businesses and approximately 47% of annual turnover in 2014. However, these companies can suffer significant financial losses as a consequence of a disruption to business operations, such as a flood event, due to their limited resources and lack of organizational plans. Agent-based modelling is recognized as one approach to enable complex problems in business and social science to be studied. SMEs' preparedness and response to disruptive events can be complex and interactive processes. Hence, agent-based modelling is an appropriate approach to study these processes and identify any emerging phenomena. With the aim of providing guidance for SMEs regarding how to better prepare and respond to the challenges faced when flooding occurs in the future, an agent-based model (ABM) is currently being designed and developed to represent and simulate SMEs' existing and potential behaviours immediately prior to, during and in the short-term aftermath of a flood event. This paper describes preliminary work on the ABM's development undertaken including the design of the various agents represented, the rules governing agent behaviours, the attributes of agents and the environment in which they operate. The basis of the ABM design draws on a range of sources including semi-structured interviews with SMEs which have experience of significant flooding, guidelines from the Environment Agency and local councils, business continuity management systems' requirements (ISO223301) and academic literature. Further, the paper discusses performance metrics of SMEs which are adopted to assess the level of continuity of business operations in the model.
Keywords: *agent-based model, business disruption, flooding, resilience, SME.*

www.witpress.com, ISSN 1755-8336 (on-line)
doi:10.2495/978-1-78466-155-7/023

1 Introduction

Small and medium enterprises (SMEs) constitute a major component of the United Kingdom's economy. Approximately, 99.9% of all private sector businesses are SMEs, accounting for 47% of private sector turnover in the UK at the start of 2014 [1]. Disruptions to businesses often result in a significant impact on the regional economy and a wider impact on the national economy. According to the Environment Agency (EA), the 2007 floods caused a £740 million loss to UK businesses [2]. SMEs are known to have limited resource [3, 4], a lack of organizational slack [5] and a lack of contingency/disaster recovery plans [6, 7]. Therefore, the capability of adapting to cope with flooding becomes vital in terms of maintaining operations.

SESAME is a research project aimed at studying and enhancing SMEs' organizational resilience to flood events [8]. The purpose of this paper is to present the preliminary design of an agent-based model (ABM), which is to be used to simulate SMEs' existing and potential behaviours immediately prior to, during and in the short-term aftermath of a flood event. The ABM offers opportunities to obtain an understanding of the resilience of SMEs at an organizational level as well as the SME community as a whole. The paper describes the design in the context of the model based on a range of sources including interviews with SMEs which have experience of significant flooding, guidelines from the EA and local councils, business continuity management systems' requirements (ISO223301) and academic literature. Further, details are presented of agents and the interaction network of agents, simulation scenario design and evaluation of SMEs' performance.

2 Related work

Limited research has been done to understand SMEs behaviours and strategies when facing floods. Previous research investigates the preparedness and adaptions of businesses in the UK when facing floods [9] and preparedness of businesses in Germany [10]. Factors identified as being related to the preparedness of a business include EA warnings, property and business insurance, resilient fittings, temporary and permanent flood installations, alternative location, alternative power source, alternative fuel source, data backup, emergency plan and business disruption plan [9]. Studies show that the majority of SMEs are inadequately prepared for floods and take a long time to recover or never reopen after floods [11, 12], while case studies in [7] provide an insight into the ability that some SMEs have in the recovery from a major flood event.

Agent-based modelling has gained increasing attention over recent years and has been applied to diverse fields such as business, biology, public policy and social science [13]. It is a tool that is commonly used to study and understand complex systems. SMEs' preparedness to flood events and recovery post flooding are complex processes. As it is difficult to analyse and model such processes using traditional approaches, agent-based modelling is seen as an appropriate means to study this problem. To date, several existing ABMs simulate either flood risk related to individuals [14, 15] or business aspects of SMEs [16, 17]. For example, Blackmore *et al.* [16] develops an ABM to understand the process of growth and innovation that evolves in SMEs over time. Another ABM [17] is developed to analyse supply chains in context of SMEs. Despite this work on ABM in the context of businesses, modelling SMEs in the context of disruptive events, in particular flooding, remains unexplored.

3 Agent-based modelling and simulation

An overview of the agent-based modelling and simulation framework is shown in Fig. 1. The framework consists of a pre-processing module, agent-based modelling and simulation and a post-processing module.

In pre-processing, the virtual geographical environment (VGE) captures all the SMEs in a defined geographical area and the dynamic flood model generates inundation data against a timeline which feeds into the ABM [18]. In the context of this paper, hydrodynamic modelling for a 1 in 1000 year event has been carried out in relation to the Lower Don Valley region of Sheffield in the UK. A 1D-ISIS and 2D-TUFLOW hydraulic model of the River Don provided by the Environment Agency has been used. More specifically, the 1D ISIS model represents the River Don channel and the 2D TUFLOW model represents the surrounding floodplain and highly developed urban areas where flood mechanisms are more complex. Buildings have been represented in 2D model domains using high roughness areas and both the 1D and 2D models are dynamically linked. The ABM consists of agents and agent networks. Various types of agents are designed in the model, whereas agent networks represent SMEs' social networks. In post-processing, the output data of the simulation are analysed to determine the level of influence of different behaviours on business operations.

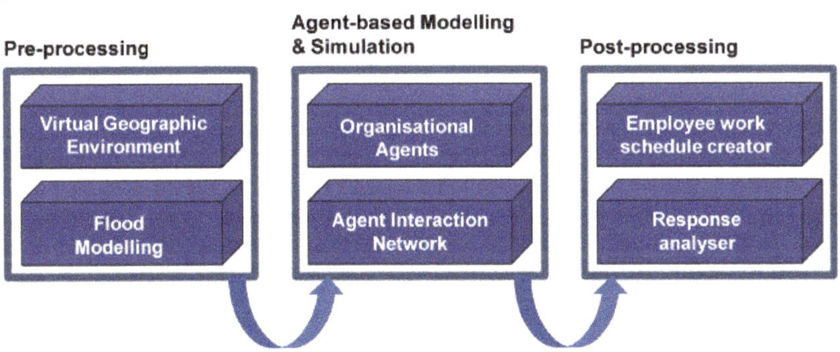

Figure 1: An overview of agent-based modelling and simulation framework.

As referred to earlier, the design of the ABM is based on interviews with stakeholders, guidelines from the EA [19] and practitioners [20, 21], business continuity management systems' requirements (ISO 223301) [22] and academic literature (e.g. [9, 10]). In this section, we will present the design of ABM.

3.1 Types of agents

To study SMEs' preparedness and responses to flood events, several types of agents have been designed and are currently included in the model:

- SME agents constitute the majority of agents in the ABM.
- Flood agents (one for each SME agent) represent the dynamic flood status in a specific geographical location, which maps to the location of the corresponding SME agents' premises.
- An Environment Agency (EA) publishes flood warnings and flood alerts, which can play an important role in flood risk management for businesses.
- A Met Office agent broadcasts weather reports and flood warnings, which if received by SME agents may accelerate their response to a flood event.

The behaviour of SME agents is influenced by the interactions between agents in the simulation environment.

3.2 Agent attributes

Agent attributes, in addition to behaviours, shape the 'personality' of an agent. Since SMEs stem from a wide range of business sectors, their attributes are categorized as generic, which all SME agents have in common, or specific, which SMEs have within a particular business sector. Some generic attributes of SME agents are obtained from the VGE output, including company name, business code according to the National Land Use Database classification, location and identification number. Other generic attributes of a business, not obtained from the VGE, include turn over, the number of employees, premises details including the number of floors and the number of buildings, and indications such as whether or not the SME owns the premises, is multi location, has an emergency-plan, has flood experience, is fully insured, is registered to receive flood warning and has cash available to aid recovery. Some other generic attributes are required to enable SME agents to communicate in networks to exchange information with other SME agents and other types of agents. These attributes include a list of neighbours as well as useful contacts such as service companies (e.g. electricians, plumbers). Apart from the generic attributes, specific attributes of SME agents vary depending on business type. Initially, this aspect of the ABM has focused on manufacturing SMEs. Specific attributes for manufacturing SMEs include stock levels and indications as to whether or not the business has engineers onsite, has an electricity generator and has pumps, all of which may aid and speed-up recovery. Other business sectors will be designed and implemented in later stages of the project.

Flood agents provide inundation data at different times in and around the premises of SME agents. A flood agent has attributes including location, and depth of water in the building and on roads adjacent to the building. The inundation at different times is modelled and calculated in the flood model and fed into the ABM. The SME agents interact with flood agents at each time step to obtain updates of the depth of water attribute and thus make decisions on which actions to take in response to the current flood situation. The EA agent and Met Office agent both have an alert table and a message sender to send/publish flood warnings and weather reports, respectively. The EA agent has a list of all the subscribers who register for the flood warnings, and only sends alerts to subscribers, whereas the Met Office agent sends weather reports and flood warnings via general media.

3.3 Agent behaviours

Behaviours of SME agents vary depending on business sector. Further, even SME agents from the same business sector can act differently in responding to a flood event according to their available resources. For example, a manufacturing SME agent could have behaviours as follows: lift up Information Technology (IT) equipment, seal doors, cut-off electricity, evacuate, contact electricians, restore power supply, hire electricity generator, contact insurance company, contact landlord, catalogue damage, clean up by employees/contractors, arrange work to be done at other unaffected locations or on floors above ground level, direct landlines to company mobiles, contact customers, contact suppliers, source money if the SME is not cash-rich, dry/fix machinery/equipment, move stock and clean stock. All behaviours are categorized into (i) pre- and during-flood stage and (ii) post-flood stage. Behaviours in the pre- and during-flood stage include saving fragile items at high risk of being damaged by the water. Behaviours in the post-flood period focus on recovery from the impact of the flood in an operational sense, e.g. working to restore production capability.

SME agents make decisions on what actions to take next based on information such as flood warnings (if received) and depth of water, leading to a prioritized sequence of behaviours to enact. Allocating a priority to each behaviour ensures that one with a higher priority will be executed first when more than one behaviour could be enacted at the same time. Also, each behaviour is considered in terms of whether or not it can be enacted before the flood water recedes from the premises, whether it can be enacted by one employee or more than one employee and whether it is generic or specific to a particular type of SME. For example, consider a manufacturing SME agent which does not subscribe to EA alerts. Such an SME may only start to lift up and move equipment when it is apparent that the flood water will enter the premises or when they watch weather reports before the water enters the premises. Also, if there are engineers on site and useful materials are available to protect the building from flooding, such as enabling doors to be sealed, these individuals can attempt to prevent water entering premises. Furthermore, if it is safe to leave the premises, evacuation would be executed. However, if roads are closed due to the inundation, employees may be stranded and would need to wait until the roads are safe to use before leaving the premises. Post flooding, the SME has to work on restoring the power supply if there is no power at the premises, clean up, inform its insurance company, contact customers and suppliers, dry and fix important machinery, move and clean stock if it is flooded, and so on.

Other types of agents have simpler behaviours compared to SME agents. Flood agents update dynamically the depth of water associated with their corresponding SME agent's premises and adjacent road. The EA agent in the model publishes flood warnings containing information of warning levels to subscribers. The Met Office agent broadcasts weather reports and flood warning to the public.

3.4 Agent networks

Case studies undertaken in the SESAME project have found that some SMEs are able to utilize internal resources, access external resources and leverage social capital forming the platform for effective response and recovery [7]. To simulate SMEs searching and accessing resources through their social networks, several networks have been considered.

- *Publish–subscribe flood warnings between the EA agent and SMEs agents*: The EA agent is responsible for publishing flood alerts and warnings. Only SMEs which subscribe to this service will receive the alert and warning messages.
- *Weather reports and flood warnings are broadcasted by the Met Office*: All SMEs are able to access information about these weather reports and flood warnings providing they are connected to media channels such as news online, television and radio.
- *Relationships with neighbours*: Neighbouring SMEs may exchange information about the flood before flooding happens and help each other during and after the flood, e.g. share warehouses unaffected by the flood.
- *Relationships with engineering services*: SMEs which have pre-existing relationship with engineering service companies (e.g. electricians and plumbers) can speed-up the resumption of operations and shorten the length of business disruption.
- *Mutual aid agreement with other SMEs*: Establishing mutual aid agreement between SMEs will help them minimize the business disruption caused by floods [22].
- *Outsourcing* [22]: SMEs disrupted by flooding may outsource their own work to other businesses unaffected by the flood thus guaranteeing continuity of the delivery of their services/products to their customers.

4 Simulation scenarios

To study SME's response to flood events, two simulation scenarios will be considered. The first scenario simulated, which is currently under development, is based on existing behaviours of SMEs with experience of flooding. The second scenario to be designed will incorporate behaviours suggested by literature as well as those identified by SMEs in light of their prior experience of flooding. Each simulation will represent a duration of 15 days commencing from the time at which inundation first occurs. An aim of these simulations is to identify the key factors and processes undertaken by SMEs which enable them to respond efficiently to a flood event, and thus accelerate recovery.

4.1 Scenario 1: SMEs with low preparedness

The first scenario to be modelled focuses on SME agents exhibiting behaviours akin to those of businesses with experience of flooding. More specifically, actions and behaviours were modelled of SMEs with experience of the significant flooding in 2007 in the Lower Don Valley region of Sheffield. Analyses of interview transcripts with a number of these businesses, which covered a range of sectors, revealed details of what they did when preparing for, responding to and recovering from the impact of a flood event.

4.2 Scenario 2: SMEs with enhanced preparedness

In the second scenario to be simulated, SME agents' behaviours will be designed based on what these businesses would do differently in terms of preparation and response, and/or in addition to what was done, as a result of their experience of flooding in 2007. Again, the behaviours identified to be modelled stem from SMEs' reflections of their own experience of flooding as indicated in the interview transcripts referred to in Section 4.1. In addition to these changes in behaviour, recommended preparatory actions for SMEs published by the EA and other government bodies will be modelled. A series of behaviour changes and recommended behaviours will be modelled including registering to receive EA flood warnings, improving the telephone system (e.g. emergency mobiles, diversion service), raising the level of equipment, stocks and documents, moving stock to other safer places, keeping sandbags on site, building bund walls, storing and backing-up all documentation electronically, preparing an emergency plan or business continuity plan, and adopting basic strategies for protecting the property such as sealing external brickwork with water proofing substances. In the simulations, both reflective behaviour changes and recommended behaviours are considered to (i) enable a comparison to be made with the results obtained from *scenario 1* and (ii) establish which factors are most influential in facilitating an effective and efficient approach to dealing with flooding, thus informing the preparedness of SMEs to future disruptive events of this nature.

5 Measurement of SMEs' performance

The current version of the ABM is aimed at enabling an investigation of the impact of organizational behaviours in relation to flooding. Evaluating the performance of SMEs and comparing the results with and without behaviour changes/recommendations allows such an investigation. However, to date no existing work has been carried out regarding the evaluation of SMEs performance when faced with a flood event. This section presents a suggested measurement of SMEs' performance after experiencing a flood event, which will be developed further as the research progresses, and an indication of the preliminary output of the model.

5.1 Performance evaluation of SMEs in the simulation

Currently, in the ABM, the impact of flooding on the performance of a SME is related to disruption time which starts from the point at which the flood causes an interruption to the SME's normal operation and ends when the company returns to normal operation. In this work, normal operation is defined as the state in which the SME has fully restored its pre-flood level capability of manufacturing products and/or delivering services. Specifically, the metric termed *operation level* (*OL*) is adopted to evaluate the performance of a SME in ABM simulations. As a result of a flood event, each SME has a list of *n* tasks $\{T_0, T_1, T_2, \dots, T_n\}$ to complete in order to return to normal operation, each of which has an associated number of man hours $\{H_0, H_1, H_2, \dots, H_n\}$. Therefore, the total number of man hours required to be undertaken in order for the SME to resume normal operation,

$$H_{total} = \sum_{i=0}^{n} H_i$$

Thus, the operation level,

$$OL = \frac{H_{completed}}{H_{total}}$$

where $H_{completed}$ is the number of completed man hours of tasks at the point in time under consideration during the recovery period. For example, consider a SME with a list of tasks which will take 550 man hours in total to complete. At the point in time when the SME's employees have worked on these tasks for 200 hours, the operation level is 36.4%, calculated from 200/550.

At any point in time, the operation level of a SME depends on the level of human resources available, along with other resources needed, and the allocation strategy of these resources to the tasks to be undertaken. Furthermore, a task can be undertaken only when there are available employees who can work on it and whether the employees have the means to carry out the task. For example, if a SME does not have power in the premises due to flooding, tasks such as clean-up and clean/fix machinery cannot be carried out, even if employees are available. As a result of the conditions and constraints to carry out tasks, the strategy that a SME adopts to allocate its human resources affects the progress of recovery. Maximizing the utilization of the human resources while considering the conditions and constraints will speed up the recovery. In addition to resources, a SME's operation level depends on the amount of disruption and/or damage caused by the flood water. In the simulations, the higher the depth of water in a premises, the more damage the it causes; thus, the more man hours needed to repair the damage, which results in longer time required to return to normal operation.

5.2 Illustrative results

An initial simulation of the ABM, focusing on manufacturing SMEs, has been set up and executed with the following assumptions:

- Manufacturing SME agents have specific attributes including stock levels, products manufactured and an indication as to whether the company has an electricity generator or not.
- Manufacturing SME agents' generic attributes are assigned within the ranges sampled from SME interview data, e.g. number of employees.
- Manufacturing SME agents have specific behaviours including dry and fix machinery, move stocks to a safer place, and clean stock.
- Some manufacturing SME agents subscribe to EA alerts and watch news and weather reports, while others do not.
- All employees of a manufacturing SME can work on tasks regardless of the type of task.

Illustrative simulation results, as processed by the post-processing module in the framework presented in Fig. 1, are shown in Fig. 2.

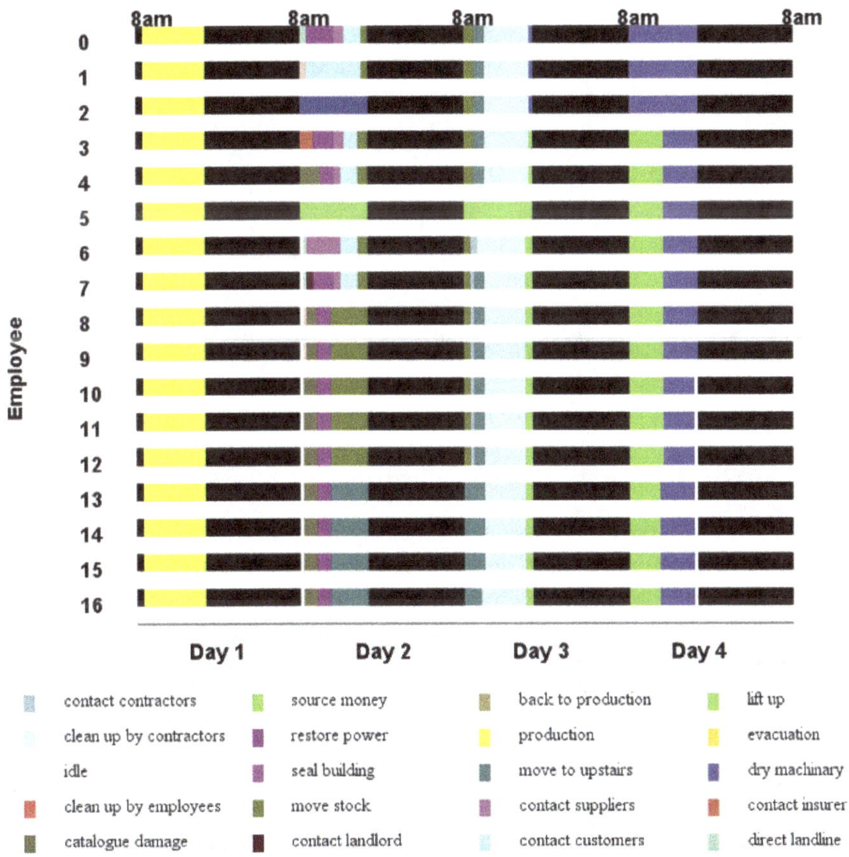

Figure 2: Employees' work schedule in a SME.

Figure 2 presents an example employees' work schedule of a manufacturing SME agent, with 17 employees. Future development of the post-processing module will include operation level being plotted against the same timeline as the employees' work schedule. As shown in Fig. 2, the SME operates from 8 am to 6 pm on each working day, with only the initial four days of the simulation shown. Employees in this SME were allocated efficiently to tasks as they are rarely idle throughout the simulation period. Note that black bars signify out of work hours (6 pm until 8 am) and it is during this period that the flood occurs (after a full working day of 'production' in the first day 1 simulated). On the second day, the SME started to allocate those employees attending work to restore the production capability. For example, *employee 0* worked on 'direct landline', 'restore power', 'contact suppliers' and 'contact customers' all of which took a relatively short time to complete due to other employees also being allocated to undertake these tasks. Among these four tasks mentioned, 'direct landline' and 'restore power' have higher priority because they must be completed before other tasks can be undertaken, i.e. dependency relationships exist between tasks. On the third day, due to there being damp stock and the 'move stock' task had not been completed by the end of the second day, *employee 0* was allocated to continue to work on this task. Further, since the premises had an upstairs level, which allowed a temporary office to be set-up, *employee 0* was allocated to 'move to upstairs' after 'move stock' was completed. Subsequently, *employee 0* waited for a contractor to clean up (i.e. 'clean up by contractors' in Fig. 2), which is arranged by the insurance company. On the fourth day, *employee 0* worked on 'dry machinery' for the entire day, after which the SME returned to normal operation. By plotting the employees' work schedule, the effect of different resource allocation strategies in the recovery period can be compared and studied.

6 Conclusion and future work

This paper presents an ABM currently under design and development, which will be used to investigate SMEs preparedness, response and recovery to flood events. Evaluation of SMEs performance in the immediate response and recovery period after floods is discussed and an illustration of how results will be presented is given. To our knowledge, this work is the first model to study SMEs preparing for and responding to flood events. The ABM will offer an opportunity for SMEs to understand how preparedness behaviours enhance flood resilience and reduce the economic impact. In future, as more business sectors will be introduced and developed in the ABM, diverse scenarios in different business sectors are to be designed and implemented to identify and investigate the key factors to improve the flood resilience of SMEs.

Acknowledgement

The authors thank the UK Engineering and Physical Science Research Council under grant EP/K012770/1.

References

[1] Department of Business, Innovation and Skills, *Business Population Estimates for the UK and Regions 2014*, Department of Business, Innovation and Skills, 2014.

[2] Chatterton, J., Viviattene, C., Morris, J., Penning-Rowsell, E.C. & Tapsell, S.M., *The Costs of the Summer 2007 Floods in England*, Environment Agency: Bristol, 2010.

[3] Van Gils, A., Management and governance in Dutch SMEs. *European Management Journal,* **23**, pp. 583–589, 2005.

[4] Sullivan-Taylor, B. & Branicki, L., Creating resilient SMEs: why one size might not fit all. *International Journal of Production Research,* 49, pp. 5565–5579, 2011.

[5] Grunert, J. & Norden, L., Bargaining power and information in SME lending. *Small Business Economics,* **39**, pp. 401–417, 2012.

[6] Jones, K. & Ingirige, B., Extreme weather events and business continuity planning. *CIB W70 International Conference in Facilities Management*, Edinburgh, 16–18 June 2008.

[7] Li, C., Coates, G., Johnson, N. & McGuinness, M., Designing an agent-based model of SMEs to assess flood response strategies and resilience. *XIII International Conference on Flood Resilience*, Zurich, Switzerland, 2015.

[8] Coates, G., Hawe., G.I., McGuinness, M., Wright, N.G., Guan, D., Harries, T. & McEwen, L., A framework for organisational operational response and strategic decision making for long term flood preparedness in urban areas. *Proceedings of the 3rd International Conference on Disaster Management*, A Coruña, Spain, 9–11 July 2013.

[9] Bhattacharya-Mis, N. & Lamond, J., An investigation of patterns of response and recovery among flood-affected businesses in the UK: a case study in Sheffield and Wakefield. *WIT Transactions on Ecology and The Environment,* **184**, pp. 163–173, 2014.

[10] H. Kreibich, I. Seifert, A. Thieken, B. Merz, D. Proverbs, C. Brebbia, *et al.*, Flood precaution and coping with floods of companies in Germany. *In Proceedings of the 1st International Conference on Flood Recovery, Innovation and Response (FRIAR), London, UK, 2–3 July 2008*, pp. 295–302.

[11] Crichton, D., Flood risk & insurance in England and Wales: are there lessons to be learned from Scotland? Benfield Hazard Research Centre: London, 2005.

[12] Crichton, D., *Climate Change and its Effects on Small Businesses in the UK*. Axa Insurance: Paris, 2006.

[13] Heath, B., Hill, R. & Ciarallo, F., A survey of agent-based modeling practices (January 1998 to July 2008). *Journal of Artificial Societies and Social Simulation*, **12**, p. 9, 2009.

[14] Dawson, R.J., Peppe, R. & Wang, M., An agent-based model for risk-based flood incident management. *Natural Hazards,* **59**, pp. 167–189, 2011.

[15] Liu, Y., Okada, N., Shen, D. & Li, S., Agent-based flood evacuation simulation of life-threatening conditions using vitae system model. *Journal of Natural Disaster Science,* **31**, pp. 69–77, 2009.

[16] Blackmore, K., Bossomaier, T., Jarratt, D. & Nesbitt, K., Intelligent agent framework for modelling the evolution of small and medium sized enterprises. *8th Australian and New Zealand Conference on Intelligent Information Systems*, 10–12 December 2003, Macquarie University, Sydney, Australia.

[17] Tounsi, J., Habchi, G., Boissière, J. & Azaiez, S., A multi-agent knowledge model for SMEs mechatronic supply chains. *Journal of Intelligent Manufacturing*, **23**, pp. 2647–2665, 2012.

[18] Coates, G., Hawe, G.I., Wright, N.G. & Ahilan, S., Agent-based modelling and inundation prediction to enable the identification of businesses affected by flooding. *Proceedings of the 4th International Conference on Flood Recovery, Innovation and Response*, Poznan, Poland, 18–20 June, 2014.

[19] Environment Agency, *Would Your Business Stay Afloat? A Guide to Preparing Your Business for Flooding*, Environment Agency.

[20] Wiltshire County Council, *Business Continuity Guide for Small Businesses,* Wiltshire County Council, 2006.

[21] Association of British Insurers. *Insurance for Small Businesses: A Guide to Protecting Your Business*, Association of British Insurers.

[22] International Organization for Standardization, *ISO 22301 Societal Security – Business Continuity Management Systems*, International Organization for Standardization: Geneva, Switzerland, 2012.

Using multi-agent technology for the distributed management of a cluster of remote sensing satellites

P. Skobelev[1], E. Simonova[1,2] & A. Zhilyaev[3]
[1]*Smart Solutions Ltd, Russia*
[2]*Department of Informatics, Samara State Aerospace University, Russia*
[3]*Department of Aircraft Engineering,*
Samara State Aerospace University, Russia.

Abstract

In recent years, the interest for space missions involving multiple satellites has been rapidly growing. This trend is responsible for an increasing need for management strategies able to coordinate the different satellites and to allocate tasks among them. In most existing systems, the management of a cluster of satellites is organized centrally and statically, which significantly reduces the responsiveness to changes in the target environment. Our research aims at designing a distributed dynamic management system where tasks are assigned to a whole constellation of satellites and their performers can adaptively change depending on emerging events. In this paper, we explore the potentiality of multi-agent technology and ontology to develop an intelligent space system. We describe our software prototype and how it can be applied to the planning problem of remote sensing satellites. In this solution, each satellite has been represented as an agent enabling communication, coordination, and negotiation among the other agents to achieve intended business goals. Moreover, we present the results of experiments showing the advantages of the proposed approach.
Keywords: distributed management, multi-agent technology, ontology, remote sensing satellite.

1 Introduction

Earth remote sensing (ERS) space systems are designed for objects and processes occurring on the Earth's surface and near space observation. The creation and application of satellite clusters is one of the promising trends in this area [1, 2]. Moreover, each spacecraft (SC) can be equipped with various equipments and can have different orbital parameters, and the solution of the sensing problem may require coordinated interaction of several spacecrafts. The management of a cluster of satellites is organized centrally and statically; each satellite has a prearranged order of task performance, set directly from the management center [2]. There is an alternative solution to implementing dynamic control, when the current tasks are given not an individual spacecraft, but to the entire cluster as a whole. This approach will allow us to reallocate tasks between cluster satellites during their implementation.

2 Problem statement

Let there be a swarm of spacecrafts capable of ground objects identification in visible, infrared, and radiation spectrum. Each SC can autonomously identify its space coordinates at any time and has devices for two-way communication with other satellites. Each SC has a cyclogram designing mechanism for its behavior scheduling. This device allows us to determine a mutual visibility with other objects. A user interacts with the swarm of satellites via the network of ground stations, where each one can transmit initial data on tasks to the swarm and receive the performance results.

A satellite cluster is given a task to dynamically identify observation areas with complex properties. In other words, it is about matching the given target object to the data obtained during observation area identification. For that purpose, we need to identify an object as one of the objects in the visible, infrared, or radiation spectrum (a combination of the spectral characteristics is possible). The observation area has the following attributes: location, priority, and a time slot when the research is needed. The same task can be solved with several satellites.

3 Methods

Using a multi-agent approach is suggested, which has been actively developing in recent years at the junction of parallel systems, distributed problem solving, artificial intelligence, and telecommunications works for dynamic task distribution implementation between spacecraft. Unlike the traditional problem-solving method that searches for a certain deterministic algorithm, allowing us to find the best solution, multi-agent technologies get the solution automatically by the interaction of multiple independent targeted software or hardware modules, that is, agents [3, 4].

3.1 Multi-agent approach to the management of a swarm of spacecrafts

In multi-agent approach implementation, a remote sensing space system is considered as a heterogeneous multi-agent system, where individual ERS satellites relay satellites that ensure rapid communication between agents in the absence of direct visibility and ground stations represent agents. In this case, a multi-agent system is a global dimensionally distributed network, where agents (nodes) either rotate around the Earth or rotate with it if the node is situated on the surface [5]. Space system management is implemented by a coordinated interaction of the network nodes. In this interaction, satellites dynamically form a team, distribute the given task between them, and solve it by parts, considering their location and on-board equipment abilities (Fig. 1).

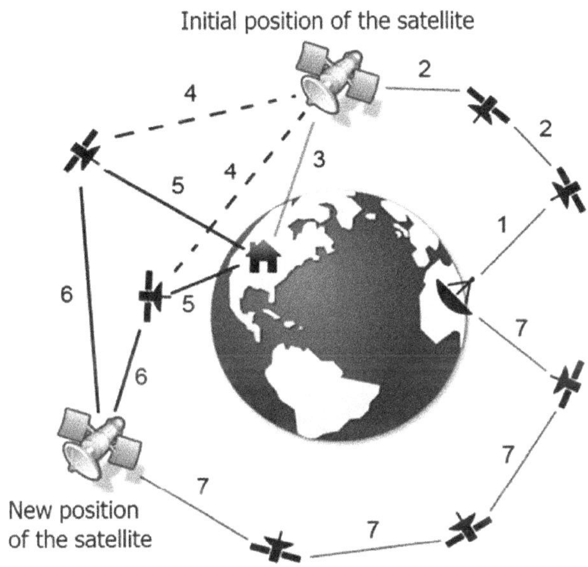

Figure 1: Network agents' interactions scheme.

1. At any moment in time, the ground station (represented by a 'radar' sign) specifies and transmits the task of a target object (represented by a 'home' sign) to search all the visible satellites (1). The tasks can be unknown to the satellites beforehand and require the cooperation of several satellites. The observation area is divided into blocks.
2. The satellites move in orbits with different characteristics. If the satellite does not 'see' the observation area, it operates as a re-translator, sending signals to the closest visible satellites. Since the satellite that received a request is far from the observation area, it transmits the request to the satellite closest to the observation area, which is in its radio availability area. The process of request transmitting continues until it reaches a satellite above the observation area (2). Among the several appropriate satellites that can identify the target, we choose the one that will fly over the observation area before the others or the one that

stays in the visibility of an observation area longer, which is determined in the negotiations of the satellite team that can see the observation area.

3. If the satellite that receives the task has the required identification equipment, it sends a message to other satellites, letting them know that it will take the task. While the satellite moves in an orbit above the observation area, it performs identification, searching for the target object in its spectrum. If the satellite does not succeed in its search and leaves the visibility area, it transmits the request to the other satellite that is entering the observation area. The process continues until a satellite identifies the object in its wave range that fits the description provided by the ground station. Let us call this satellite an initiator (3).
4. Simultaneously, the satellite-initiator transmits the request of area research to other satellites operating in a different wavelength range (4).
5. Satellites of these ranges check the specified object by the coordinates transmitted by the satellite-initiator (5).
6. Satellites send the research results to the satellite-initiator directly or through the re-translators network (6).
7. If the research result is positive, then the satellite-initiator sends a message with it to the ground station. After that, the satellite-initiator continues the search for other objects that fit the description given by the ground station in the observation area (7).
8. The scanning is finished with the ground station signal or when all of the observation area blocks are searched by the equipped satellites that work in all the wavelength ranges.

A multi-agent approach application allows us to adaptively reallocate the tasks inside the swarm by the interaction between the spacecrafts. Moreover, in this type of system, it becomes possible to process the unscheduled events, such as new top-priority task arrival, spacecraft failure, placing new spacecraft into the orbit, etc. Spacecrafts can not only independently and individually make decisions, but also negotiate and flexibly form satellite alliances (or teams) of different purpose if necessary. At the same time, the signal that initiates rescheduling of the whole swarm can be sent both from the Earth's surface and from any of the satellites, for example, from the one that was first to discover the target object.

3.2 Ontological description of the target object

The aim of an ERS spacecraft swarm is to establish a match between the map of the observation area and a specified pattern that describes the ratio of certain basic elements such as a house or a road between them. An ontology implemented as a semantic network is used for a target object description formalization. An ontology is a conceptual domain knowledge represented in a computer-readable format and applied in decision-making. Ontology conceptual knowledge means that the knowledge is formulated in terms of basic concepts (the most common entities and relations) that describe domain area fragments [6]. A semantic network is a directed graph, where nodes represent some concepts and the arcs are the relations between them.

A target object description is sent from the ground station in the form of a tree of connections, where the terminal nodes (leaves) are basic concepts, nonterminal nodes are composite concepts, and the arcs are the connections between them. Generally, the tree can contain concepts of different types of spectrum. A composite concept can be made of the basic ones with the relations 'not', 'or', 'center', 'near', 'before', etc. An example of an ontological description of a target object in a tree form is shown in Fig. 2.

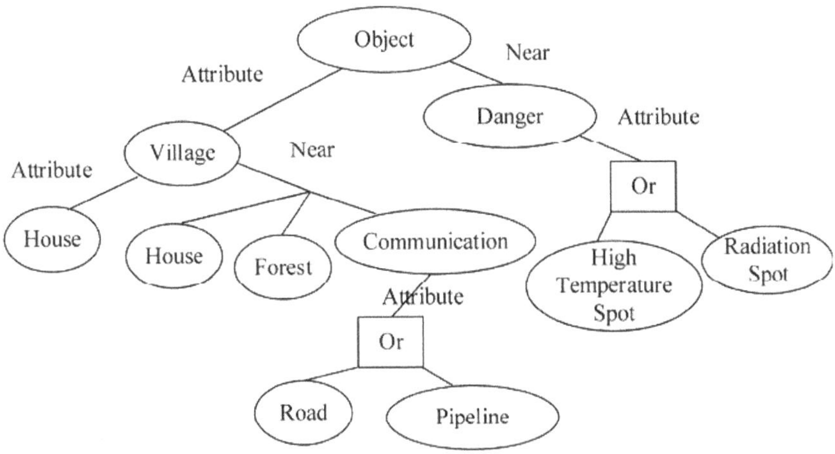

Figure 2: Ontological description of a target object.

According to the description in Fig. 2, the target object is a house in a village that has the additional attributes of high temperature or radiation level. The 'village' concept is defined as a group of houses with a forest, a road, and a pipeline nearby.

A spacecraft that received a request for the observation area sensing extracts a subtree from the whole tree, containing only those concepts that are in its visibility area. The spacecraft performs a sensing according to its ontological description subtree and in the case of a positive outcome, sends a request to spacecraft for sensing in other spectra, transmitting a corresponding subtree along with the task description.

A spacecraft of a corresponding visibility spectrum can explore each subtree of the ontological description of a target object regardless of the spacecraft of the other spectra, except when the different spectra concepts are connected by the 'or' or 'not' relations. For example, if an SC of a radiation spectrum explored the observation area and did not find a high radiation level, then it cannot make a conclusion that the observation area does not fit the ontological description, because it may contain a high temperature object.

Ontologies application allows us to formalize specific domain knowledge in a computer-readable format and separate this knowledge from the system software code. This creates a basis for further system development and the addition of new features without constant reprogramming.

4 Results

4.1 Simulation system

A simulation system for distributed management research methods of moving objects in the satellite clusters was developed (Fig. 3). An ontology creation module and dynamic scheduling module that form a processing plan of the flow of incoming events (observation area sensing request, equipment failure, etc.) and perform adaptive dynamic change of the created plan are the main components of the simulation system.

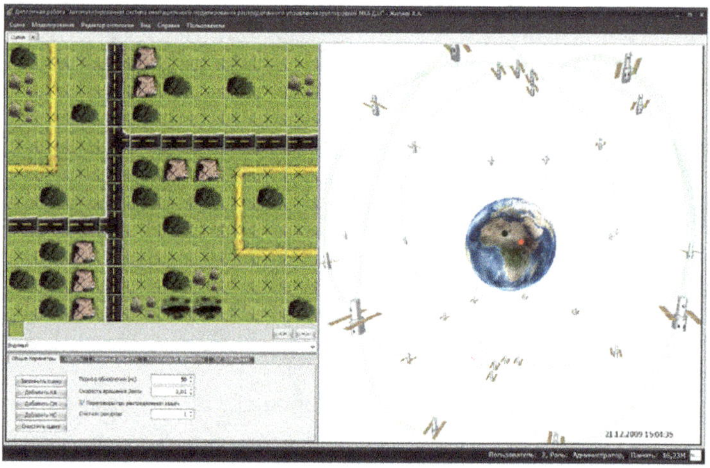

Figure 3: Simulation system screen.

4.2 Experimental research

Let us use the following satellites cluster model. Let there be 32 spacecrafts with different spectrum equipments and three relay satellites that are situated on the geostationary orbit in the ERS. The task is the sensing of several closely situated observation areas, whose number in different series of experiments can change from one to six.

Two groups of 30 tests are carried out during the experiment:

1. new tasks are centrally allocated to the most appropriate SC without considering their current loading (batch scheduling) and
2. rescheduling is performed when a new order arrives, by the means of interaction (negotiations) between the satellites (real-time scheduling).

Moreover, numerous tests are divided into six series that differ in the number of observation areas in the scene.

Table 1 shows the time values (in terms of simulated time) spent on the observation areas' exploration and that are obtained during the experiment. The values of arithmetic mean and standard deviation of the exploration time of the observation areas are shown for each series of tests.

Table 1: Comparison of sensing time.

Number of observation areas		1	2	3	4	5	6
x	Batch scheduling	42.8	57.4	66.6	103	103.8	130.8
	Real-time scheduling	43.6	57.8	59.4	86.8	90.6	117.2
σ	Batch scheduling	2.79	2.5	1.6	8.97	18.08	25.48
	Real-time scheduling	3.07	1.72	2.5	3.76	3.61	13.47

A graphical representation of the experimental results is shown in Fig. 4.

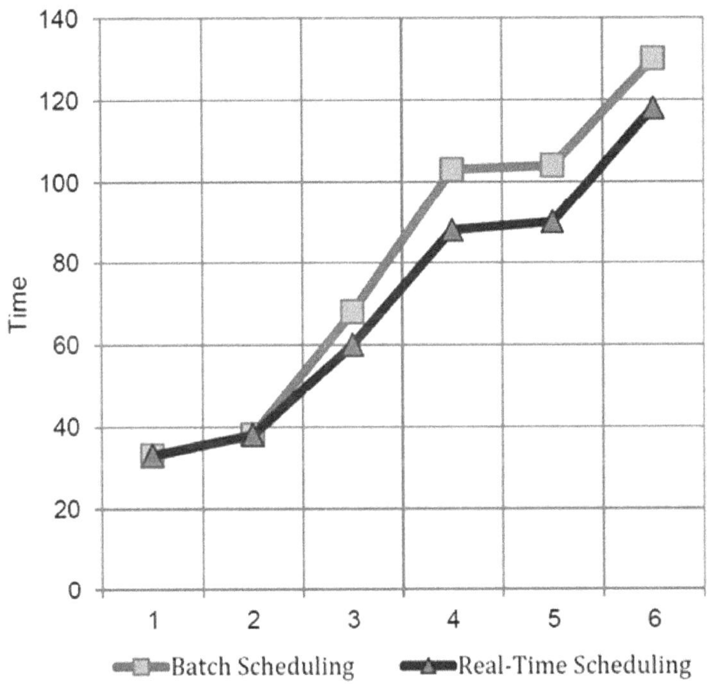

Figure 4: The average sensing time chart.

Using agents' negotiations allows us to significantly decrease the standard deviation of the amount of time spent on the observation areas sensing. Therefore, total time of the ERS tasks execution becomes more definite and is easier to predict.

Let us consider the experiment in the ERS tasks execution time analysis under the condition of several spacecrafts leaving the swarm. Let us use the example of sensing of four observation areas that are evenly distributed along the Earth's equator. The number of areas and their locations remain the same throughout all the experimental series. Several spacecrafts are dynamically eliminated from the swarm. The number of eliminated spacecrafts increases in each next series of the experiment. The selection criterion of the eliminated satellite is its mandatory participation in the sensing process of one of the observation areas. For each experiment, the result will be total time of the observation areas sensing by the resources of the rest of the satellite swarm.

Table 2 shows the time values (in terms of simulated time) spent on the observation areas sensing and that are obtained in the course of the experiment. A graphical representation of the experimental results is shown in Fig. 5.

Table 2: Observation areas sensing time, under the condition of spacecraft elimination.

Series no.	Observation areas sensing rime, in simulated time terms, considering eliminated spacecraft							
	0	1	2	3	4	5	6	7
1	72	114	110	142	130	165	197	364
2	80	95	118	126	169	174	238	411
3	69	92	129	138	134	168	226	376
4	70	101	115	131	126	189	245	382
5	81	117	134	122	151	175	223	447
x	74.4	103.8	121.2	131.8	142	174.2	225.8	396

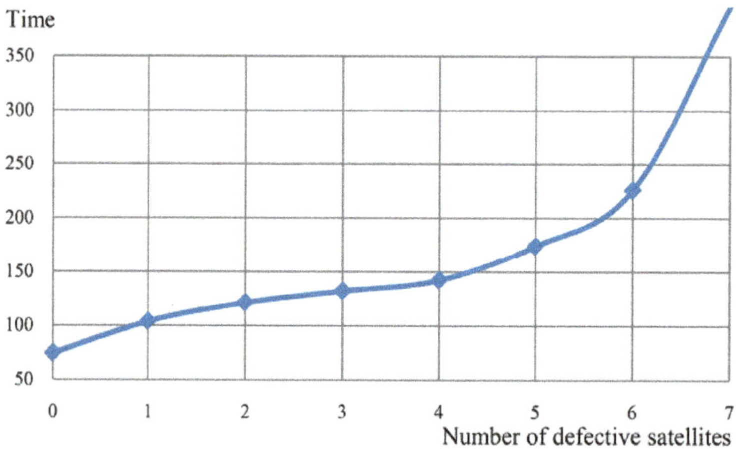

Figure 5: Average sensing time under the condition of SC elimination.

Comparison of the sensing duration in an experimental series that has a different number of eliminated spacecrafts has shown that with each SC elimination, its tasks are redistributed among other spacecrafts of the swarm. Moreover, the sensing task solution takes more time with the new conditions, as the new executors watch the observation area in less time than the eliminated ones.

5 Conclusion

The experiments conducted have shown the potential of the suggested approach to the swarm of spacecraft management in terms of complex dynamic environments. The solutions must not only be optimal in a global sense, but also appropriate and fast in the current situation. Preliminary results show that adaptive pattern recognition and scheduling generate results of higher quality faster and cheaper than in previous arrangements. In addition, the solution is scalable and reliable.

Acknowledgements

This work was carried-out in SEC 'Smart Solutions' Ltd with the financial support of the Ministry of Education and Science of the Russian Federation (Contract no. 14.576.21.0012, unique number RFMEFI57614X0012). P. Skobelev was the project manager and E. Simonova the senior analyst.

References

[1] Tripp, H. & Palmer, P., Stigmergy based behavioural coordination for satellite clusters, Acta Astronautica journal, 66, pp. 1052–1071, 2010.
[2] Iacopino, C., Palmer, P., Policella N., Donati A., Brewer A. How Ants Can Manage Your Satellites, Acta Futura, 9, pp. 57–70, 2014.
[3] Wooldridge, M., An Introduction to Multiagent Systems, John Wiley & Sons: London, 2nd edition, 2009.
[4] Rzevski, G. & Skobelev, P., Managing complexity, WIT Press: London-Boston, 1st edition, 2014.
[5] Sollogub, A., Intelligent System for Distributed Problem Solving in Cluster of Small Satellites for Earth Remote Sensing, Information and Control Systems, eds. P. Skobelev, E. Simonova, A. Tzarev., M. Stepanov, A. Zhilyaev, 1(62), pp. 16–26, 2013.
[6] Skobelev, P., Activity ontology for situational management of enterprises in real time, Ontology of designing, 1(3), pp. 6–38, 2012.

Multi-agent method to adaptive real-time train scheduling with conflict limitations

A. A. Belousov[1,2], A. A. Goryachev[2], P. O. Skobelev[3] & M. E. Stepanov[1,2]
[1]*Samara State Aerospace University, (National Research University), Russia*
[2]*SEC "Smart Solutions", Russia*
[3]*Group of companies "Knowledge Genesis", Russia*

Abstract

The paper proposes a multi-agent method to creating an intelligent adaptive system for train real-time scheduling with conflict limitations. The architecture of the multi-agent system consisting of two base planning subsystems is described. Subsystem interaction protocols and protocols of agent interaction within each subsystem are presented. An example of scheduling in various situations is presented in detail. Characteristics of the developed multi-agent system are presented. The high quality of scheduling process and system performance is shown.
Keywords: multi-agent technology, intelligent systems, interactions method, real-time planning, conflict solving.

1 Introduction

It is widely known that distribution of different types of transport can be very different in transportation systems of regions and separate countries of the world. Thus, transportation system of industrialized countries has a complex structure and is represented by all types of existing transport including electronical. Such countries as Japan, USA, France, Germany, and Britain have a transport infrastructure with an especially high level of development. In our days, 85 percent of the world's internal cargo turnover belongs to the industrialized countries (without seagoing transportation). In Western Europe 25 percent of the cargo turnover belongs to the railway transport, 40 percent to motor transport and the

WIT Transactions on State of the Art in Science and Engineering, Vol 90, © 2016 WIT Press
www.witpress.com, ISSN 1755-8336 (on-line)
doi:10.2495/978-1-78466-155-7/025

remaining 35 percent, to inland water-borne transport, maritime short-range cabotage and pipelines [9]. In the USA about 10 percent of cargo turnover belongs to the railway transport [10].

Talking about mainland part of the industrialized countries, the portion of 10-25 percent of turnover, without passenger traffic, is critical and will continue to attract attention of economists in the near future. As the construction of railway infrastructure continues since the beginning of the 19th century, its present condition each time leaves less and less variants for extensional development and pushes towards the intensive development: usage of energy-saving technologies and traffic modes, improvement of infrastructure services and, for the most part, train traffic schedule optimization.

Today, due to the high traffic intensity, trains are highly interconnected: changes in one train's schedule, or a conflict with this train, will affect neighboring trains and can have impact on the whole train network in a very unpredictable way. In this case, rescheduling of all trains in the operating domain might be required, which should be done rapidly, in real-time adaptive manner, and this is a very complicated task, given the whole variety of planning conditions, preferences and limitations as well as security requirements.

In spite of the high level of railway network automation, solving conflict situations, which is a complicated multi-objective task, is still carried out solely by dispatchers. These professionals are not limited in usage of PC capacity, however, at the same time they are considerably limited in tools that could help them in performing their main function: building optimal nonconflictual schedules of train traffic. Demand for such automation systems is growing with the rapid increase of railway traffic density up to critical values and is especially visible in stress situations of deviations from the initial plan when decisions have to be made very quickly.

Modern smart systems for real-time decision-making support need to be based on new methods and tools of management automation, which can take into account a whole variety of factors, criteria, rules and interactions. Moreover, the influence of each factor is quite individual and can be changed during operation. Still, such systems provide a very high level of precision and quality of decisions/results as well as productivity.

This paper describes the multi-agent approach, which solves the complex task of adaptive real-time train management. At present, the developed system is in industrial implementation in dispatcher control center of October Railway, a part of JSC RZD.

2 Multi-agent approach to adaptive train route management

2.1 Problem statement

The task of adaptive real-time railway management consists in creating a detailed train route plan with minimal deviations from the master-plan. Building of such a route plan, various limitations and requirements for continuously incoming events must be considered (occupations of infrastructure block sections, track maintenance, train damage, speed limitations, etc.).

The solution of this task is the optimal schedule that brings all the affected trains back to the master-plan as soon as possible. The best method is real-time rescheduling which solves all the existing and foreseen conflict situations.

The input data includes the following: detailed railway infrastructure (stations, station-to-station intervals, railway switches, block sections), requirements for train schedule and maintenance, updates on the current situation in the operating domain (trains and block section status).

The process of scheduling faces many different formalized (normative route-building requirements, traffic security requirements, dispatcher rules) and hard-to-formalize limitations (no unjustified halts or trains stays on the main tracks, no unjustified changing of tracks, no unjustified high density of trains between stations, etc.).

Besides, implementation of any kind of requirements and decision-making depends on the current situation.

For instance, a train may choose the wrong track if there is a busy block section or a maintenance window on its path, in order to bypass the obstacle and stay on time. Such activity can be acceptable only if it does not affect the other trains with higher priorities. Another option is to stay on the same track and wait, which is possible if the delay is short and the train can later catch up. But for staying at a station the train may only choose block sections of a certain length. In any case, the decision whether the route should be changed or not is linked to many conditions which require finding balance and consensus.

For solving the task of adaptive real-time train route management there has been developed a system based on multi-agent methods and technologies. This multi-agent approach takes into consideration numerous formalized conditions, restrictions, abilities and hard-to-formalize requirements during negotiation among agents which aim to find the consensus. Due to the open architecture of the multi-agent system it is possible to input new conditions and criteria as well as change the already existing ones and manage them dynamically.

The multi-agent system for solving the task of adaptive real-time train route management runs on Vektor-M program platform [1, 2]. This system allows keeping the dynamic infrastructure model of the operating domain, receiving signals from block sections, appointed maintenance windows, satellite positioning data and other information.

2.2 Architecture of the multi-agent system

The architecture of the developed system is built on the network-centric principles, where every subsystem has its own individual task and the final solution is composed of separate solutions reached by different agent swarms via negotiation [3].

The primary plan building is performed in 2 subsystems. Each subsystem builds a train route schedule on its own level of understanding of the scene in such a way that the initial rough decision is transformed into a more precise one. A decision made in every subsystem is conflict-free for its level of understanding (no converging train routes, security requirements are intact). This layer-based train scheduling eliminates the combinatorial explosion of possibilities, making the scheduling process more stable to disruptions due to reducing the scale of the task on higher levels and considering all the possible limitations according to the level of importance and impact on other layers.

All events occurring in the system can be divided into two main types: new request or update on the current situation. Requests in their turn can be of two basic types: request to let the train pass or request to conduct maintenance works. Updates on the current situations can be either a train moving along block sections or a state of an infrastructure (damage or busy condition).

The first level of planning is represented by a path scheduler, the second one - by a time scheduler. A general decision-making method and the role of schedulers in it are depicted in Fig. 1.

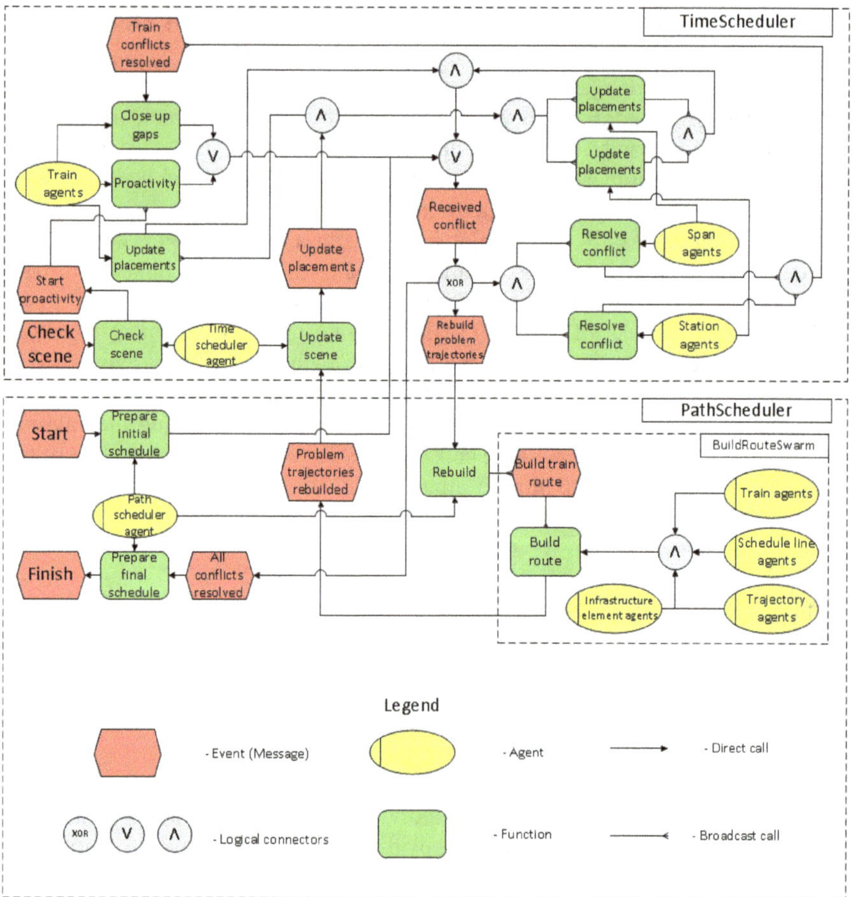

Figure 1: Interaction between main planning subsystems, represented by a diagram of the Event-driven Process Chain.

The diagram describes the interaction between systems on a high level. The diagram blocks in their turn can be subsystems with their own complex inner structure.

Work in the system starts with a message "Start", which is sent to the agent "Path scheduler agent". In response, the function "Prepare initial schedule" appears, which prepares the planning scene and after which the event "Received conflict" appears in the time scheduler. As a result, the time scheduler detects the current schedule conflicts and determines further direction. If no conflicts are found, "Path scheduler agent" receives the message "All conflicts resolved", and later the function "Prepare final schedule" forms the overall schedule. If conflicts are found, there are two scenarios: to send the message "Rebuild problem paths" to the path scheduler or to solve the conflict on its own level, by changing time of resource occupation and sending "Resolve conflict" to station and station limits agents.

In the first case, the path scheduler will change tracks and planned block sections of train stops in the schedule problem zone and will send a message "Problem paths are rebuilt" to "Time scheduler agent"; then the "Scene update agent" is called, which launches a new chain of rescheduling and negotiations in the demand-request network between requirements of train traffic and infrastructure elements, after which the event "Received conflict" emerges again in the time scheduler.

In the second case, in the process of resolving conflicts in their own schedule, station and station limits agents create conflicts in train schedule, so called "Gap", which appear due to the difference in arrival and departure time on the adjacent elements of infrastructure. Solving "Gap" conflicts takes place in the method "Close up gap", implemented by train agents, resulting in new conflicts of resource occupancy, which is checked when the event "Received conflict" appears. Reaching a compromise between train agents and agents of infrastructure elements (stations and station limits) is the final objective of the scheduling system.

If a compromise is reached and all conflicts are resolved, but the decision does not satisfy the requirements, a message "Check scene" initializes the process of proactivity, and consequently, certain agents try to improve their own schedule and cause a new chain of resolving conflicts.

Time scheduler represents an operating domain as a set of station limits and stops, which builds a train schedule in less details, conflicts are resolved by queueing the trains, speeding them up and slowing them down.

Time planning is similar to a visual schedule analyzing. The main task is to build a new possible train schedule considering the normative schedule limitations and train priorities. The solution is based on the method of Demand-Resource Networks for managing resource allocation in real time [4]. On this level a train agent creates subtasks (operations) for passing a station limit or staying at a station for a certain amount of time. An agent of every subtask of this kind looks for a placement for itself in the respective resource, trying to find the most profitable position by negotiating with other subtask agents. High-priority trains are more active in finding a placement (they have more power for pushing out other requests for resources).

The main decision-making condition here is accomplishing the task with minimal divergence:

$$D = \sum_{i=1}^{X} \left(\left| ST_s - AT_s \right| + \left| ST_f - AT_f \right| \right) \rightarrow \min \qquad (1)$$

where STs – scheduled starting time, STf – scheduled finishing time, ATs – actual starting time, ATf – actual finishing time, X – number of resources (station limits, station platforms), where operations of passing and stopping can be implemented. Additional conditions implemented while making decisions are listed below. The result of time scheduling is a schedule for stays and passes of stations and station limits in an operating domain, which is sent to the path scheduler in order to build train routes according to block sections, if it is possible.

During scheduling in the time scheduler there are two types of interaction:
1) interaction between interval agents of different trains on one block section: this algorithm supports providing for an order change of moving trains on block sections;
2) interaction between interval agents of one train on different block sections. This algorithm enables keeping the path continuity of each train (if a train is late at some block section, it requires correcting the schedule of this train on all its block sections).

Each type of interaction is used for decision-making of respective type of conflict, and one type of conflict might result in appearance of conflicts of other types.

In the path subsystem a train route is built according to the graphics which are computed in less detail, conflicts are resolved due to bypassing and route changing.

The primary task of scheduling paths and stays is to allocate routes for trains and choose the block section for their stays. In this subsystem a train agent creates new subtask agents, which look for routes for passing the station and station limits according to the condition of minimal route costs. "Cost" is the cumulative key performance indicator of the route, which includes different normative requirements for train routes (correct or not, length, number of connections). After scheduling with minimal KPIs, station route agents enter the active phase, where the main condition for decision-making is no overlapping of train routes in a block section. When such an overlap is found, a station route agent will try to transmit one of the conflicting subtasks to other route agents. Route agents communicate via the task exchange protocol [5, 6].

Building and negotiating the final train schedule takes place in close interaction between scheduler levels. In every planning subsystem there is a swarm of agents, representing this level, moreover, between the subsystems there are back links which come into play when a conflict cannot be resolved locally in the current subsystem. The primary allocation of tasks to resources is done based on the best decision possible independently and in concurrent threads, which allows for reduction of computing time by excluding the rest of possibilities. Such "greedy" allocation leads to conflicts that are resolved by agents grouping together into structures within a swarm – domains. In each domain searching for a compromise takes place between agents in order to resolve the conflict.

2.3 Criteria for schedule quality assessment

Setting optimization criteria is based on several requirements imposed on the final schedule. These requirements were collected from system users, and the main ones are listed below. For convenience, all the criteria are grouped.

Group "Initial planning"

1) No unjustified deviations from master-plan. Deviations can be justified by:

- Inconsistency between schedule and infrastructure (no available stay at station, no 4.5-minute train interval spacing, forced passing loop on a single-track road, noncompliance of bulk cargo graphics, etc.).
- Impact of disruptive situations (windows, train delay/advance, busy condition, speeding down or train priority, etc.).

Group "Conflicting traffic organization"

2) No unjustified choice of the wrong (conflicting) track. Choosing the wrong (conflicting) track can be justified by:

- non-bypassing window on the right track or bypassing window(s), which can cause delay more than 4 minutes (breaking/acceleration average time);
- fixing the train on a wrong track or block sections, so that only switching to the wrong track is possible.

Group "Stays planning"

3) No unjustified train stays. Justified stays are:

- stays justified by master-plan;
- stays caused by a passing loop on a single-track line;
- stays caused by trains with lower priorities giving way to the ones with high priority;
- stays caused by bypassing of obstacles;
- 24-hour stays caused by speeding down of trains.

4) No train stays on tracks with not enough length.

5) No stays of heavy-loaded trains on "heavy profile track".

6) No long stays on the main tracks (for instance, suburban train stays for 1 minute).

7) Sustaining the given intervals of non-simultaneous train arrivals to a station.

Group "Routing of dispatcher schedule trains"

8) Routing of dispatcher schedule trains should not lead to deviations of other trains, except for trains with the same priority as dispatcher schedule train.

9) While routing of dispatcher schedule trains there should be no unjustified stays. Example of a justified stay is giving way to passenger trains. Theoretically, arrival should pass the operating domain without stays at stations.

Agents need to communicate with each other in order to be able to comply with requirements. Any communication creates a special network that satisfies the current demands of system performance. Simplification of the semantic network between agents of the proposed system is shown in Fig. 2.

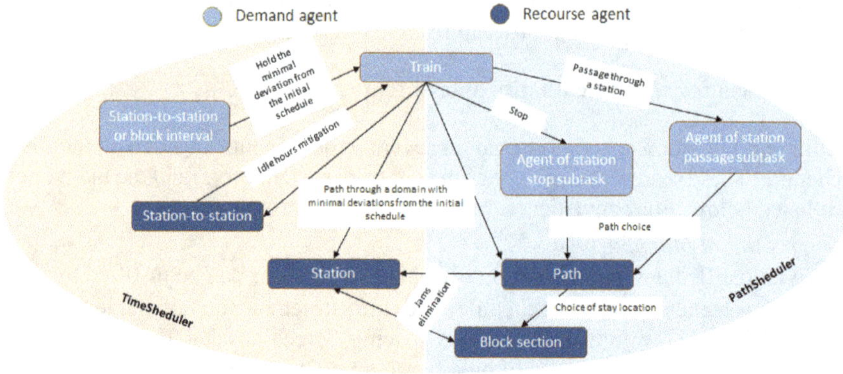

Figure 2: Simple chart of the virtual world semantics.

3 Software implementation

3.1 Implementation features

For performance analysis of the multi-agent system for adaptive real-time train route management the following time indices were used: infrastructure load time, train load time in an operating domain, rescheduling time for a new arrival, rescheduling time depending on duration of maintenance window, rescheduling time for added maintenance windows, rescheduling time depending on the number of tracks with maintenance windows, rescheduling time depending on speed limits on a block section, rescheduling time depending on the number of limitations, speed on a block section, rescheduling time depending on the number of tracks occupied due to a speed limit at a station.

A train schedule in operating domain has around 45 operations (operations of passing, stays at a station or passing a station) and every train operation has its own agent. There are around 800 trains in total, so the total amount of agents is around 36000. Apart from that, there are train agents – around 800, station agents – 49, station route agents – 500, block section agents – 3700, maintenance request and availability agents – around 100–200. Dividing this many agents into levels and grouping them into isolated swarms of agents, which are active at certain points of time, allows for a significant increase in performance [3].

Average values have been determined for each index, based on the results of 10 experiments for two operating domains: Saint-Petersburg–Buslovskaya and Saint-Petersburg–Moscow.

Figures relevant for planning the features of the operating domains are represented in Table 1.

Table 1: Relevant figures of operating domains.

Operating domain	Number of stations	Number of infrastructure objects	Number of railway switches	Average load and preparation time of infrastructure (ms)
Saint-Petersburg–Buslovskaya	17	1293	133	1452
Saint-Petersburg–Moscow	47	3640	304	7194

The Moscow–Saint-Petersburg operating domain has 2.8 times more infrastructure objects than the Saint-Petersburg–Buslovskaya operating domain. The load time of the Moscow–Saint-Petersburg operating domain is about 4.9 times higher than on the other operating domain. It is caused by building of the infrastructure model within the system and special infrastructure station agents, station limits, block sections.

Table 2 represents the change of scheduling characteristics depending on the number of scheduled trains on the operating domains.

Table 2: Scheduling characteristics depending on the number of tasks.

Operating domain	Number of trains	Average scheduling time (ms)	Average rescheduling time (ms)
Saint-Petersburg–Buslovskaya	12	1511	1185
	17	2180	1429
	30	2257	1799
	49	2300	2092
	67	2746	2356
Saint-Petersburg–Moscow	71	7202	5063
	86	8786	6164
	152	12691	11420
	241	22453	22899
	311	36143	37351

According to Table 2, the difference between the scheduling times on the operating domains with around the same number of trains is 2.6 times, which is close to the difference between domain infrastructures. Thus, dependency can be observed, which is close to linear, between the number of infrastructure elements and the scheduling time of trains on an operating domain.

Rescheduling time is 15–40% less in comparison with the initial scheduling time. It is caused by adaptive rescheduling based on incoming events instead of rescheduling from scratch. Increasing the number of trains increases the train scheduling time in direct proportion.

Table 3 represents scheduling characteristics depending on the density of disruptions. Increasing density increases the scheduling time.

Table 3: Scheduling characteristics depending on the density of disruptions.

Operating domain	Density of disruptions	Average scheduling time (ms)
Saint-Petersburg–Buslovskaya	No disruptions	2356
	Very few disruptions	2476
	Few disruptions of the same type	2505
	Many disruptions of different types	2785
Saint-Petersburg–Moscow	No disruptions	37351
	Very few disruptions	39631
	Few disruptions of the same type	37613
	Many disruptions of different types	39501

The scheduling time mostly depends on the number of infrastructure elements and the number of scheduled trains. Other factors such as disruptions, their number, duration and density have less influence on the total scheduling time. This can be explained by partial (adaptive) scheduling according to incoming disruptive events instead of complete rescheduling. Thus, one can speak about guaranteed time for decision-making in the given infrastructure and given number of trains.

The following qualitative characteristics can be noted: no violations of the given train arrival intervals, no unjustified changing of tracks, no traffic jams between stations, keeping security intervals, almost no delays among intercity and high-speed trains in conflict situations, average train delays less than 9% (20 trains are in one conflict). This result has been achieved on such large-scale planning tasks for the first time.

3.2 Example of resolved conflict situations

Let us consider an example of how the system deals with conflict situations and take a closer look at the decisions made by the scheduler. The situation with a high number of disruptions is shown in Fig. 3. There are 5 maintenance windows in both directions.

In order to resolve this situation it is required to involve back links between different planning levels. Due to a high density of graphic lines between the windows the path subsystem was unable to create the final schedule, since the traffic security requirements didn't allow all trains to stop and switch on to the alternative route. The path subsystem registers this inconsistency as a conflict and sends a message to the time subsystem. In order to resolve the situation, the time subsystem must delay a few trains from the previous stations, taking the overload of the station limits into account. It sends the newly made decision to the path subsystem, which builds the route and checks whether the decision satisfies traffic security requirements. While resolving all the conflicts the system also takes into consideration train priorities, time for braking and acceleration. Dramatic schedule changes have also been prevented. As a result, the schedule became more balanced and stable to possible further disruptions, the effect of the maintenance window

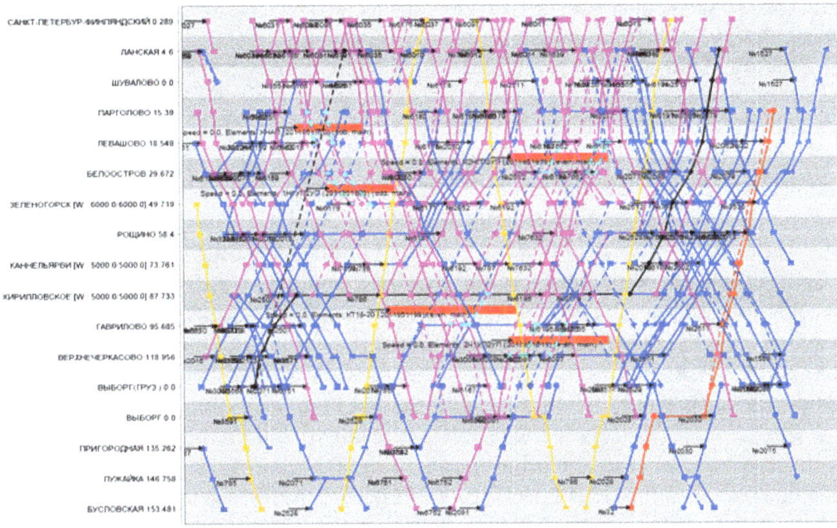

Figure 3: Bypassing 5 maintenance windows.

has been localized in an hour, after which the schedules tend to be more regular (high density was decreased).

An overload between even two stations cannot be resolved without the time subsystem, because changing the route with switching on to the opposite tracks (for double-sided route) will cost much more than changing the train schedule.

3.3 Implementation

The expected results of the developed adaptive train scheduling system include: reduced reaction time, increased flexibility and quickness of decision-making in response to disruptive events, increased efficiency of railway resource management in real time and securing on-time performance of trains, reduced man-hours for rescheduling trains, a completely new intelligent software system for traffic management in real time.

At present, there is no detailed statistics comparing the results of scheduling system and dispatchers. However, it is beyond any doubt that even the most qualified dispatchers react to unpredictable events slower than the system, which has higher quality level of the final schedule and compliance with the strict requirements.

For instance, train routing with dispatcher schedule represents a complex task, considering the whole variety of affecting factors (dimensions, weight, length in carriages, type of load, normative movement, braking and acceleration time, etc.). It is preferable for a train to pass an operating domain as quickly as possible and with minimal number of stays, but at the same time, without any conflicts with other trains. Routing a cargo train with bulk load is quite a complex task because it cannot cross with high-speed trains due to security requirements. Although an experienced dispatcher can easily handle this task even with dense schedule, it will take him two or more minutes.

Routing of several trains from different directions under conditions of technical works on a station limit requires correct calculating of all standards for speeding up and down while setting on a stay, choosing alternative tracks, all speed limitations, all track occupancy at a station, all lengths of free stays and so on. One should also remember about customary trains which pass an operating domain and different unpredictable events happen. In this situation, a dispatcher needs more time for decision-making, at least a couple of minutes for each train.

In fact, the schedule does not remain static, and due to external events might become impossible in several seconds. For example, busy condition of a track of required length at a station due to damage, makes the planned train stay impossible. Deviations from the master-plan worsen dispatcher's statistics, and hence affect his wages, so this situation results in greater attention of dispatcher and can get lower priority to the problem of qualitative train routing.

In case of delay of a high priority train, it should be put back in the master-plan as quickly as possible. This is a critical and higher-priority task in comparison with others. A dispatcher, acting according to the situation and taking into account low priority of bulk trains, can put it on a timely stay. At the same time, every unplanned stay of a cargo train causes unpredictable costs and has a negative effect on dispatcher's statistics.

Hence, a dispatcher has to control a great number of indicators and at the same time take into account his wages. However, thoroughly thinking over low-priority tasks is not always possible. In its turn, a scheduler, in spite of the variety of events, disruptions, conditions and limitations, can manage the train routing task, including dispatcher schedule, in seconds and gives a dispatcher a decision which satisfies all the limitations and requirements, with minimal deviations from master-plan and minimal number of stays (thus, reducing energy expenditures).

4 Conclusion

The suggested multi-agent system for adaptive real-time train scheduling [3] has been developed within the project of the smart trains scheduling system for the Russian railways [7, 8].

The scheduler development is still in progress and the next steps will include the following: logics track monitor for dispatchers, decision explanation map, improving interaction with the dispatcher, modelling of foreseen "bottle necks" and signaling about them, self-learning, scoring the quality of the decision, increasing performance by paralleling asynchronous planning processes.

Acknowledgement

Work is performed with financial support of the Ministry of Education and Science of the Russian Federation (the agreement No. 14.576.21.0014, unique identification number RFMEFT57614X0014).

References

[1] A.B. Shabunin, S.N. Markov, D.V.Dmitriev, N.A. Kuznetsov, P.O. Skobelev, S.S. Kozhevnikov, E.B. Simonova, A.V. Tsarev. Integrative platform for the network-centric approach to creating allocated intelligent resource management systems at Russian Railways, Software engineering, 2012, Issue 9, pp. 23–28.

[2] A.B. Shabunin, N.A. Kuznetsov, P.O. Skobelev, I.O. Babanin, S.S. Kozevnikov, E.V. Simonova, M.E. Stepanov, A.V. Tsarev. Developing an ontology for the multi-agent resource management system for Russian railways, Information technologies, 2012, Issue 12, pp. 42–45.

[3] A.A. Belousov, P.O. Skobelev, G.A. Efremov, M.E. Stepanov, A.A. Goryachev, A.B. Shabunin. Multi-agent approach to a complex task of train scheduling in a large-scale railway management systems

[4] Vittich V.A., Skobelev P.O. Method of Demand-Resource Networks for managing resource allocation in real time, Autometria, 2009, 2nd issue, pp. 78–87.

[5] Abbink, E.J.W., Mobach, D.G.A., Fioole, P.J., Kroon, L.G., van der Heijden, E.H.T., Wijngaards, N.J.E., Actor-Agent Application for Train Driver Rescheduling, In: Proceedings of AAMAS, Budapest, Hungary, 2009, pp. 513–520.

[6] P.O. Skobelev, A.A. Belousov, S.O. Lisicin, A.V. Tcarev. Development of the intelligent cargo logistics management system for Vostochny polygon, Papers of XV World Conference “Problems of management and simulation in complex systems”, Samara, 25–28 June 2013, Samara: Samara Science Centre RAS, 2013, pp. 391–396, ISBN 978-5-93424-662-5.

[7] A.B. Shabunin, N.A. Kuznetsov, P.O. Skobelev, I.O. Babanin, S.S. Kozevnikov, E.V. Simonova, M.E. Stepanov, A.V. Tsarev. Developing a multi-agent resource management system for Russian railways, Mechatronics, automation, management, 2013, Issue 1, pp. 23–29.

[8] Matuchin V.G., Shabunin A.B.IRMS. Concept and implementation // Papers from the first conference “Intelligent railway management systems IRMS-2012”, Moscow, 15–16 November 2012, pp. 15–18.

[9] World transport. Transport meaning in world economy. Types of transport and their peculiarities. Transport and environment. [online] Available at: <http://geographyofrussia.com/transport-mira-znachenie-transporta-v-mirovom-xozyajstve-vidy-transporta-i-ix-osobennosti-transport-i-okruzhayushhaya-sreda/>

[10] Freight facts and figures: U.S. Department of Transportation, 2013. [online] Available at: <http://www.ops.fhwa.dot.gov/freight/freight_analysis/nat_freight_stats/docs/13factsfigures/pdfs/fff2013_highres.pdf >

Author index

www.ingramcontent.com/pod-product-compliance
Lightning Source LLC
LaVergne TN
LVHW011847200926
848484LV00002B/414